T0304518

Mathematical Modeling

Branching Beyond Calculus

TEXTBOOKS in MATHEMATICS

Series Editors: Al Boggess and Ken Rosen

PUBLISHED TITLES

PUBLISHED TITLES CONTINUED

PUBLISHED TITLES CONTINUED

Mathematical Modeling

Branching Beyond Calculus

Crista Arangala

Nicholas S. Luke

Karen A. Yokley

CRC Press
Taylor & Francis Group
Boca Raton   London   New York

CRC Press is an imprint of the
Taylor & Francis Group, an **informa** business

A CHAPMAN & HALL BOOK

CRC Press
Taylor & Francis Group
6000 Broken Sound Parkway NW, Suite 300
Boca Raton, FL 33487-2742

© 2018 by Taylor & Francis Group, LLC
CRC Press is an imprint of Taylor & Francis Group, an Informa business

No claim to original U.S. Government works

Printed on acid-free paper
Version Date: 20171219

International Standard Book Number-13: 978-1-4987-7071-2 (Hardback)

Visit the Taylor & Francis Web site at
http://www.taylorandfrancis.com

and the CRC Press Web site at
http://www.crcpress.com

Contents

Preface

This textbook is intended for use in a second or third year undergraduate course directed toward mathematics and other STEM majors. The book is divided into chapters based primarily on prerequisite structure and is designed to be flexible to what previous classes are required and the goals of the modeling course. Each chapter begins with a list of goals and expectations in order to help instructors know the different expected prerequisite knowledge and find the sections that best align with their individual objectives.

Each section focuses on a mathematical topic from calculus or later in a traditional curriculum or a topic related to model analysis. Within each section is a short list of closed-end exercises, and the examples and these exercises are intended to expose students to how different applications can be quantified and analyzed. Additionally, each section contains projects, which guide students through more in-depth investigation. These projects can be used as in-class directed activities or out-of-class assignments. Ideally, students will also be involved in the creation of models, and a "Chapter Synthesis" section is included with each chapter with more open-ended problems or exercises that direct students to revisit previous problems with different perspectives. The first chapter of the textbook is intended to emphasize the discussion-based nature of modeling and includes activities focused on discussing what answers mean and what can be learned from visual representations. Mathematical models are excellent tools to use to learn about the physical world. However, mathematicians and modelers must be able to understand results in context in order to gain insight. The first chapter is intended to guide students (with backgrounds in calculus) to expand their communication skills and their perspectives.

Subsequent chapters focus on particular mathematical content and how that content can be used with applications. The textbook is intended as a resource that can be tailored based on the prerequisites and goals of the particular course. For example, a course that has calculus II and linear algebra prerequisite courses may want to use material solely from Chapters 1-3. Another instructor who teaches a course with a technology emphasis or computer science prerequisite may take a mix of material from Chapters 1-4 and heavily use Chapter 5. A course that focuses on the modeling process (with more student autonomy in how the models are created) could be designed around the Chapter Synthesis sections. While the prerequisite expectations are discussed at the beginning of each section, some prerequisites may not be as crucial depending on the use of technology in the course. Chapter 4 is based on previous knowledge of differential equations, but topics could be included in a course without a differential equations prerequisite depending on the classroom introduction and technological use. Throughout the textbook, exercises and projects marked with a TE have a technological emphasis. The use of mathematical software will facilitate the solving of these problems.

The creation of this textbook was supported in part by Elon University FR&D funding.

1

Modeling Concepts, Visualization, and Interpretation

Goals and Expectations

The following chapter is written toward students who have completed one semester of collegiate calculus.

Goals:

- Section 1.1 (Modeling and Representation): To understand how equations and functions can be used to quantify basic scenarios.

- Section 1.2 (Articulation and Interpretation Using Calculus):

  1. To review calculus concepts related to graphical shape.

  2. To strengthen students' articulation of calculus concepts connected to applications.

- Section 1.3 (Transformations of Functions):

  1. To investigate and review how specific changes in algebraic function definitions affect graphical output.

  2. To strengthen understanding of how to change a function to achieve a particular result.

- Section 1.4 (Sensitivity): To introduce the idea of model sensitivity to parameters.

1.1 Modeling and Representation

Mathematical modeling is a broad subject. Essentially, when a non-mathematical scenario is described using an equation or equations or by a set of mathematical rules, those equations or rules are called mathematical models. Creating a mathematical model involves understanding how to quantify events or changes, and using a model requires understanding both the mathematics and how the representation (or model) is defined.

Strong mathematical knowledge is necessary to form conclusions when mathematics is applied to other subjects. Additionally, one must be able to communicate results both in mathematical terms and in the language of the application. This section will focus on considering functions as representations and articulating information related to those functions.

First, recall that a function, $f(x)$, takes an input, x, from the function's *domain* and matches that input with an output, y, from the function's *range*. No single x value will result in more than one matching output y. A common way of checking if a graph is a function is the *vertical line test*, where one knows that a graph represents a function if no vertical line can be drawn that crosses the graph in more than one place. When a function, $f(x)$, is also a model, that function takes some representative input and matches it with a related output. Both the input and output will be quantifications in some way, either a measurement, count, percentage, or other numerical representation.

Example 1.1.1 *Consider the function $A(t) = 100e^{-\frac{\ln(2)}{3}t}$ where A is the amount of a particular decaying chemical in grams after t years. The initial amount of the chemical present is 100 g and the half-life of the chemical is 3 years. A graph of $A(t)$ is shown in Figure 1.1. Consider the following:*

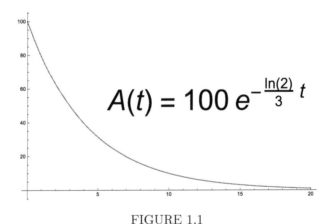

FIGURE 1.1

1. *What is the amount of chemical present after 4 years? Round your answer to 2 decimal places.*

2. *When (in years) will the amount of the chemical be 25 g?*

3. *When (approximately in months) will the amount of the chemical be 90 g? Round your answer to one decimal place.*

In part 1, the question is asking what A is when $t = 4$. We can answer this by inserting $t = 4$ into the equation.

$$A(4) = 100e^{-\frac{\ln(2)}{3}(4)} \approx 31.50.$$

After 5 years, approximately 31.50 g of the chemical are present.

In part 2, the question is asking what is t when $A = 25$.

$$
\begin{aligned}
25 &= 100e^{-\frac{\ln(2)}{3}t} \\
\frac{1}{4} &= e^{-\frac{\ln(2)}{3}t} \\
\ln\left(\frac{1}{4}\right) &= -\frac{\ln(2)}{3}t \\
-\ln(4) &= -\frac{\ln(2)}{3}t \\
\frac{3\ln(2^2)}{\ln(2)} = \frac{6\ln(2)}{\ln(2)} &= 6.
\end{aligned}
$$

In 6 years, 25 g of the chemical will be present.

Note that we could have reasoned out the answer without the function $A(t)$ based on other information in the given question. After one half-life period (or 3 years), 50 g of the chemical should remain. After another half-life period, 25 g of the chemical will remain.

Finally, in part 3, the question is asking to find a t value given a particular value for A. However, part 3 asks for the t answer to be presented in different units than the function uses.

$$
\begin{aligned}
90 &= 100e^{-\frac{\ln(2)}{3}t} \\
\ln\left(\frac{9}{10}\right) &= -\frac{\ln(2)}{3}t \\
-\frac{3\ln\left(\frac{9}{10}\right)}{\ln(2)} &= t.
\end{aligned}
$$

We will want to round our answer, but we also need to convert the answer to be in months. Hence,

$$t_{months} = 12\left(-\frac{3\ln\left(\frac{9}{10}\right)}{\ln(2)}\right)$$

$$\approx 5.5\,.$$

The sample of the chemical will be reduced to 90 g in approximately 5.5 months.
∎

   In the process of Example 1.1.1, we needed to pay attention to several aspects of the problem. Understanding what is an input and what is an output is important to each part of the problem. This example is a fairly simple function and involves only one independent variable, but connecting notation correctly to quantities being described is very important in all mathematical models. In other mathematical texts, the focus of a problem like Example 1.1.1 might be to construct the function $A(t)$ (and the reader may want to try this for practice). The problem was presented differently in the previous example in order to view mathematical models the way they are often presented. The example presented the equation then explained what it described. The person using the equation has a responsibility to investigate the model enough to be confident that the model describes the application appropriately. The answer to Example 1.1.1, part 2 could be found without an equation modeling the chemical amount over time, but checking the validity of the model is a good idea. Additionally, problems based on applications often require mathematical flexibility. Unit measurements are often indicative of how a model is structured, and one should always be conscious of units in application problems.

Example 1.1.2 *Consider the function $P(t) = 30t^2 - \frac{16t^3}{3} + \frac{t^4}{4}$ where P is the yearly profit (in thousands of dollars) of a toy producer t years after the first product is available for sale. A graph of $P(t)$ is shown in Figure 1.2.*

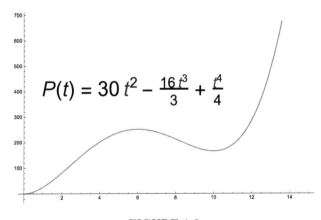

FIGURE 1.2

1. *What is the estimated yearly profit at the end of 2 years after the first prod-uct is available? Present your answer rounded to the nearest penny.*

2. *What is the estimated yearly profit at the beginning of the 11th year of the products being available? Present your answer rounded to the nearest penny.*

3. *Using a calculator or computer software, determine the approximate time when the company's yearly profit was $50,000. Round your answer (in years after the first product is available) to two decimal places.*

The first question is asking for the value of P given a particular value of t. The problem states that t is the number of years after the product is available for sale. Therefore, when asked about the yearly profit at the end of 2 years, we should find the yearly profit when $t = 2$. We can calculate $P(2)$

$$P(2) = 30(2)^2 - \frac{16(2)^3}{3} + \frac{(2)^4}{4} = \frac{244}{3}$$

but we need to remember that the output, P, is in thousands of dollars. Hence, we need to convert to dollars to answer the question.

$$\frac{244}{3} \cdot 1000 \approx 81333.33.$$

The estimated yearly profit of the company at the end of the 2 years after the product is available is about $81,333.33.

Now, the second question is asking for the value of P given a particular value of t. However, the time is presented at the beginning of a year as opposed to the end of one. One year begins at the instant the previous year ends. Therefore, the problem is asking for the estimated yearly profit after 10 years have passed or when $t = 10$.

$$P(10) = 30(10)^2 - \frac{16(10)^3}{3} + \frac{(10)^4}{4} = \frac{500}{3}$$

$$\Rightarrow \quad \text{Profit in dollars} = \frac{500}{3} \cdot 1000 \approx 166666.67.$$

The estimated yearly profit of the company at the beginning of the 11th year is about $166,666.67.

Finally, the third question is asking to find a particular t value given a particular value of P. Because P is in thousands of dollars, we want t when $P = 50$. In other words, we want to find the solution(s) to

$$50 = 30t^2 - \frac{16t^3}{3} + \frac{t^4}{4}.$$

Two real values exist where $P(t) = 50$ but only one is positive. Solving the above equation for $t > 0$ results in $t \approx 1.49$. Hence, the company's estimated yearly profit is \$50,000 when the product has been available for approximately 1.49 years.

∎

Note that the function $P(t)$ does not describe total profit over time. Similar to the previous example, in Example 1.1.2, knowing what values go with t and which go with P are key to solving the problems. Additionally, we have less information about the problem in order to sense if the model is reasonable. For example, the company is always making a profit in Example 1.1.2, which may not be representative of many companies. Additionally, we have no information about a period of time that the model represents. Making predictions in time is difficult and a model may only be reasonable for a limited period of time. Let's consider the same function from Example 1.1.2 but in the context that the function is based on yearly profit figures for the first 10 years after products are available for purchase. $P(t)$ from Example 1.1.2 is replotted with data points at the end of the first 10 years in Figure 1.3. The trend of the graph after $t = 10$

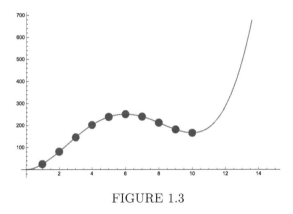

FIGURE 1.3

is very different than for $0 < t < 10$. Knowing the limitations of any model is important when using that model to make conclusions about the application described.

Exercises

1. Consider the profit of a company

$$P(x) = x^2 - 4x - 21$$

where P is the total profit of the company in millions and x is the number of products manufactured (in thousands). How many products does the company need to produce in order to make a profit?

2. A petri dish initially contains 50 bacteria, and these particular bacteria double in number every hour. A person in the lab says that the same petri dish contains 150 bacteria after 2 hours have passed.

   (a) Explain, using calculations based on initial bacterial numbers and doubling times, why you think the person in the lab is incorrect.

   (b) Assume you have double checked the person's bacterial count and the dish did, in fact, contain 150 bacteria after 2 hours. Explain how your math in part 2a can be correct, but the petri dish contains a different number. (In other words, what are you assuming with your calculations that may not be true?)

3. Consider the model presented in Example 1.1.1:

$$A(t) = 100e^{-\frac{\ln(2)}{3}t}$$

   where A describes the amount of a chemical (in grams) with half-life 3 years t years after an initial amount of 100 g is present.

   (a) You are interested in finding out how much of the chemical remains after 10 years. Using only what you know about the half-life and the initial amount of chemical, find two numbers A_1 and A_2 such that $A_1 < A(10) < A_2$.

   (b) Use the formula to find $A(10)$ rounded to two decimal places. Verify that your answer is between the two numbers you found in part 3a.

4. Consider the model presented in Example 1.1.2:

$$P(t) = 30t^2 - \frac{16t^3}{3} + \frac{t^4}{4}$$

   where P is the yearly profit (in thousands of dollars) of a toy producer t years after the first product is available for sale. However, assume this model is representative of the first 10 years after the toy is available. Find what the model $P(t)$ predicts for the profit of the toy producer 11 and 12 years after the first product is available to the nearest dollar. Also state what those values mean (in a complete sentence each) in the context of the problem. Also explain which is likely a better prediction and why.

5. Consider the model

$$R(x) = \frac{25x}{x+7}$$

where R is the rate of metabolism of a particular enzyme in mg/h as a function of the concentration of the chemical being metabolized in mg/L.

(a) What is the maximum rate of metabolism of the enzyme?

(b) What is the rate of metabolism of the enzyme when the chemical is present in a concentration of 40 mg/L?

(c) What concentration of chemical present results in half the maximum metabolic rate?

6. The North American elephant populations are not self-sustaining owing to poor reproduction. Poor reproductive rates are largely due to the limited number of bulls in captivity. The zoo that you work for would like to bring in some bull (male) elephants to mate with their females; however, they ask you to determine the age of the bulls that they should bring in. You determine that the probability of fertility in bull elephants can be modeled by the function

$$p(t) = \frac{80000}{t^{\frac{3}{2}}} e^{-\left(\frac{30}{t} + t - 2\right)}$$

where t is the age of the elephant in years. Use your model for probability of fertility to determine whether it would be better for the zoo to bring in a bull elephant that is between the ages of 3 and 4 years old or between the ages of 4 and 5 years old. (The bull will only be brought in for a short period of time.)

Project: Exponential Modeling

A linear relationship occurs when a 1 unit increase in input results in a fixed addition or subtraction in output. We call this fixed addition (which may be positive or negative) the *slope* of the line. An exponential relationship occurs when a 1 unit increase in input results in a fixed factor or multiplied change in output.

For the following project, consider bacteria placed in a petri dish. The dish initially has 100 bacteria and adequate resources. This particular type of bacteria reproduces every 5 minutes; hence, the population total doubles every 5 minutes.

1. How many bacteria are present in the petri dish after 5 minutes? After 10

minutes?

2. Explain (in complete sentences) how the number of bacteria in the petri dish involves an exponential relationship (as described above)?

3. Form a function $P(t)$ that describes the number of bacteria, P, in the dish after t minutes (assuming unlimited resources).

4. Determine (approximately) how many bacteria are in the dish after 1 hour. Then determine how many bacteria are in the dish after 1 hour and 1 minute. Use your answers to estimate the factor by which the total population changes at each step.

5. Use time t and $t + 1$ and properties of exponents to verify your estimate in part 4.

6. Is using this model as prediction for the number of bacteria in the dish reasonable to use for any t? What physically would happen once t gets large? What happens to the model when t gets large? Fully explain both what you would expect to happen and what the model currently predicts.

Project: Modeling Lens Folding

Consider a large, lightweight mirror or lens intended to be used in space. Such an object is restricted by how much it can be bent or folded and still function properly. Hence, a lens of this type must be carefully handled in order not to damage its structure. However, in order to transport a large lens into space, the lens must be compacted into a cylindrical rocket. This project will focus on one model for folding the lens and how that model is used to determine if folding the lens in a particular way is reasonable. The large mirror being considered is a doubly curved membrane, much like a very large contact lens, with a hole in the center. The lens will have an overall diameter, D, and a smaller diameter, d, will define the hole in the center of the lens, as shown in Figure 1.4. D and d will depend on the purpose of the lens and can be considered fixed measurements. More information on the application and the model discussed in this project can be found in [1].

The lens itself may be cut into pieces, but fewer cuts are preferable in order to have as much functionality as possible. One way to potentially package the lens is to make one cut along a radius of the overall lens and then fold it into roughly a cone shape (ignoring the existing curvature of the lens). (Note: in [1], this configuration is referred to as the "Single Cut Model.") The simplified cone

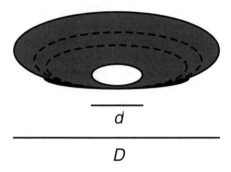

d

D

FIGURE 1.4

is shown in Figure 1.5. The cone would be formed by overlapping the lens upon itself, and as a result, the radius of the hole at the top of the cone would not be the same as the radius of the hole of the lens itself (or d). Using the geometry of a cone, we can make recommendations as to whether this folding plan is reasonable. As shown in Figure 1.5, the radius of the top of the cone is denoted R_t (for "top" radius), and the radius at the base of the cone is denoted R_b. In order to make decisions about what might damage the lens, the "bending" of the lens must be quantified. The most tightly distorted area of the lens will be the location of the greatest curvature (assuming the lens is bent *with* its existing curvature and not against), which will be the location with the smallest radius. $R_t < R_b$; therefore, we can consider how much the lens will be distorted by investigating R_t. However, the lens must still fit within a rocket in order to be deployed, which we will consider when thinking about reasonable values for R_b.

1. Assume that we know the radius of the rocket, R_{rocket}, and in order to constrain the lens within the rocket, set $R_b = R_{rocket}$. Use similar triangles to form an equation connecting D, d, R_t, and R_{rocket}.

2. Let $0 < c < 1$ be the fraction the diameter of the hole is to the overall lens. In other words, let $d = cD$. Use this new definition to form an equation that connects R_t, R_{rocket}, and c. Solve the equation for R_t.

3. The rocket will have a fixed size, and the center radius, d, will likely depend on what is needed from the lens. Let the rocket radius be 2 m, and let $c = \frac{1}{2}$. Under these conditions, what is R_t? If the lens designer said that the radius could only be as small as 50 cm and not distort the mirror, would you suggest folding the lens into the cone shape?

4. If the rocket radius is 2 m and the smallest radius of the folding could not

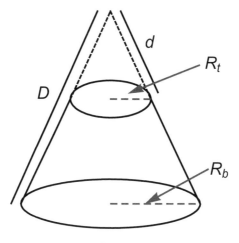

FIGURE 1.5

be less than 50 cm, what would c need to be in order to use the cone folding scheme effectively? Does your answer make sense with what we would expect for the overall problem? What would you expect?

5. All models are simplifications of more complicated situations. As demonstrated in part 4, a model may not capture every aspect necessary to effectively predict values of interest. Describe ways the cone model could be adapted to better describe the shape of the lens after making one cut along an overall radius. In other words, describe what aspects of the shape were ignored or oversimplified and how those aspects could be more fully incorporated. Your discussion can include qualitative descriptions or equations or both. You may consider shapes other than the cone.

1.2   Articulation and Interpretation Using Calculus

Many mathematical models are based on concepts from calculus (at some level). In addition to having a foundational knowledge in calculus, one must also be able to articulate this understanding in order to appropriately use calculus-based models. Additionally, more intricate models are often evaluated based on graphical results, and interpretations can be made based on visual output. This section will focus on a review of calculus concepts related to graphical shape of functions and how to describe these aspects.

Recall for a differentiable function $f(x)$:

- $f(x)$ is *increasing* when $f'(x) > 0$.

- $f(x)$ is *decreasing* when $f'(x) < 0$.

Note that when we describe function behavior, we use words such as "when," "on," and "at" to connect the behavior of the function at particular locations. Hence, we use input or x values with this particular language.

When a function is described as *increasing* on an interval, the value of $f(x)$ is becoming larger as x becomes larger within the x values of the interval. Conversely, a function is described as *decreasing* when the value of $f(x)$ is becoming smaller as x becomes larger. Consider the cubic function $f(x) = x^3 - 3x$. We can take the derivative of this function, $f'(x) = 3x^2 - 3$, and determine that

(i) $f(x) = x^3 - 3x$ is increasing on $(-\infty, -1) \cup (1, \infty)$, and

(ii) $f(x) = x^3 - 3x$ is decreasing on $(-1, 1)$.

This function is shown in Figure 1.6 and we can see how the graph reflects (i)-(ii) above. We can also see in the graph that $f(x) = x^3 - 3x$ has one "peak" and one "valley" where the function changes from one behavior to another.

Further recall for a function $f(x)$:

- A number b (in the domain of $f(x)$) is a critical number of $f(x)$ if either $f'(b) = 0$ or $f'(b)$ is undefined.

- $f(x)$ has a relative maximum at c if $f(c) > f(x)$ for all x near c.

- $f(x)$ has a relative minimum at d if $f(d) < f(x)$ for all x near d.

- $f(x)$ can only have relative extrema at critical numbers.

One way to determine the location(s) of relative extrema is to find all critical numbers of a function and then determine the behavior (increasing or decreasing) on either side of each critical number. This method, known as the *First Derivative Test* uses the following classification:

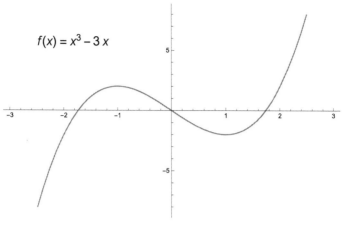

$f(x) = x^3 - 3x$

FIGURE 1.6

- If $f(x)$ is increasing (or $f'(x) > 0$) before b and decreasing (or $f'(x) < 0$) after b, then $f(x)$ has a relative maximum at b.

- If $f(x)$ is decreasing (or $f'(x) < 0$) before b and increasing (or $f'(x) > 0$) after b, then $f(x)$ has a relative minimum at b.

- If $f'(x)$ does not have different signs on opposite sides of b, $f(x)$ has neither a maximum nor a minimum at b.

As shown in Figure 1.6 and described in (i)-(ii), $f(x) = x^3 - 3x$ changes from increasing to decreasing at -1 and changes from decreasing to increasing at 1. We can add to our previous list to include

(iii) $f(x) = x^3 - 3x$ has critical numbers -1 and 1,

(iv) $f(x) = x^3 - 3x$ has a relative maximum at -1, and

(v) $f(x) = x^3 - 3x$ has a relative minimum at 1.

However, the characteristics (i)-(v) do not uniquely define the specific function we are investigating. We can use calculus to characterize additional graphical aspects of a function. We can return to the idea of increasing and decreasing but now consider the behavior of the derivative of our function of interest. For a twice differentiable function $f(x)$:

- $f'(x)$ is increasing when $f''(x) > 0$.

- $f'(x)$ is decreasing when $f''(x) < 0$.

The list above is exactly the same as presented earlier in this section except one prime has been added to each function. The derivative of a differentiable function, $f(x)$, is a function itself, and we can use the information about $f'(x)$ to understand what $f(x)$ looks like. If $f'(x)$ is increasing, then the values of $f'(x)$ are getting bigger as x gets bigger. In other words, as we move to the right in the graph, the slopes of the tangent lines to $f(x)$ are becoming larger. If these slopes are positive, the graph will become steeper. If these slopes are negative, the graph will become flatter. The resulting shape is similar to a smile or upward bowl, and the graph is referred to as *concave up* on intervals where $f'(x)$ is increasing.

If $f'(x)$ is decreasing, the slopes of the tangent lines to $f(x)$ are becoming smaller. (Note that "becoming smaller" may mean getting more negative.) In the case where $f'(x)$ is decreasing, the graph will become flatter if the slopes are positive and the graph will become steeper if the slopes are negative. The resulting shape is similar to a frown or upside-down bowl, and the graph is referred to as *concave down* on intervals where $f'(x)$ is decreasing. The abbreviated conditions for concavity for a twice differentiable function $f(x)$ are summarized below:

- $f(x)$ is *concave up* when $f''(x) > 0$.

- $f(x)$ is *concave down* when $f''(x) < 0$.

Additionally, we say that a function $f(x)$ has an *inflection point* at a number in its domain if the function changes concavity at that number.

The function $f(x) = x^3 - 3x$ has second derivative $f''(x) = 6x$, which is positive for positive x and negative for negative x. Hence,

(vi) $f(x) = x^3 - 3x$ is concave up on $(-\infty,0)$,

(vii) $f(x) = x^3 - 3x$ is concave down on $(0,\infty)$, and

(viii) $f(x) = x^3 - 3x$ has an inflection point at 0.

In addition to information based on derivatives, a function can also be characterized by trends as x approaches infinity or negative infinity, which is sometimes referred to as the function's *end behavior*. Recall for a function $f(x)$:

- If $\lim_{x \to \infty} f(x) = L_1$, $f(x)$ has the horizontal asymptote $y = L_1$.

- If $\lim_{x \to -\infty} f(x) = L_2$, $f(x)$ has the horizontal asymptote $y = L_2$.

Note that a function could have the same asymptotic behavior on both the left and right sides of the graph (in other words, L_1 could equal L_2), or could have completely different behavior on the different "ends." A graph could have no horizontal asymptotes, one horizontal asymptote (on only one or on both sides), or (at most) two horizontal asymptotes.

The function $f(x) = x^3 - 3x$ has no horizontal asymptotes, but we can still characterize the graph using calculus related to end behavior:

(ix) $\lim_{x \to \infty} x^3 - 3x = \infty$ (or $f(x) = x^3 - 3x$ increases without bound as $x \to \infty$)

(x) $\displaystyle\lim_{x \to -\infty} x^3 - 3x = -\infty$

The items (i)-(x) significantly describe the graph of the function $f(x) = x^3 - 3x$ which we see in Figure 1.6. Although a thorough examination of the function and its derivatives would be ideal, descriptive aspects similar to (i)-(x) could be roughly determined from graphs (if graphical information is all that is available). Additionally, understanding how these characteristics relate to the first and second derivatives of a function can be helpful when making interpretations about rates of change in application.

Example 1.2.1 *Consider the function $g(x)$ on $[0,\infty]$ (presented in Figure 1.7) and the following questions.*

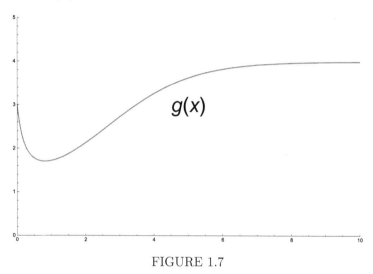

FIGURE 1.7

1. *Estimate the intervals where $g(x)$ is increasing and decreasing.*

2. *Estimate where $g(x)$ has any relative extrema.*

3. *Estimate the intervals where $g(x)$ is concave up or concave down.*

4. *What (if any) are the horizontal asymptotes of $g(x)$?*

1. The function $g(x)$ appears to be decreasing on $(0,0.8)$ and increasing on $(0.8,\infty)$.

2. The function $g(x)$ appears to have a relative minimum at 0.8.

3. The function $g(x)$ appears to change from concave up to concave down somewhere close to 3. (The actual value is closer to 2.7, but estimations will be limited by graphical scales.) Therefore, $g(x)$ appears to be concave up on about (0,3) and concave down on (3,∞).

4. The function $g(x)$ appears to have a horizontal asymptote of $y = 4$, and this behavior occurs as $x \to \infty$.

■

Note that in Example 1.2.1, all answers were presented in sentences. While all these questions could be answered more briefly, questions related to applications require more discussion. Describing graphs should be viewed similarly and will serve as better preparation for transitioning to questions that require interpretation.

Example 1.2.2 *Consider the function $g(x)$ on [0,10] (presented in Figure 1.7) as representing the average rating on a website (out of 5 stars) of a movie x months after its release date. Assume viewers can continue to vote on the website throughout the 10 months plotted.*

1. *When is the public's rating of the movie improving? When is rating declining?*

2. *When is the average rating of the movie the lowest?*

3. *When is the public's attitude toward the movie (as measured by this particular website) improving at the fastest rate?*

4. *Based on the information presented in the graph related to the first 10 months after the movie's release date, what do you expect the average rating to be a year after the movie has been released?*

1. The movie's popularity appears to decline (beginning at a value of 3 out of 5 stars at its release date) until almost the end of the first month. Near the end of the first month after the movie's release, popularity begins to improve and improves throughout the rest of the 10 month period.

2. The average rating of the movie is the lowest near the end of the first month after its release.

3. The rating is improving once the movie has been released for a little under one month (roughly after $t = 0.8$). The steepest place shown on the graph after $t = 0.8$ occurs near $t = 3$ or around the end of the third month. The rate of change of the movie's rating is at a relative (and absolute) maximum here. Therefore, the public's attitude toward the movie is increasing at the highest rate at about the end of the third month after the movie's release.

4. The trend of the average rating appears to settle to about 4 out of 5 stars once the movie has been released for a few months. The expected rating would probably be about 4 out of 5 stars after the movie had been released for 1 year.

∎

Note that Example 1.2.1 and Example 1.2.2 present extremely similar questions with respect to content. (However, Example 1.2.2, part 3 deals more with the inflection point specifically rather than the intervals of concavity.) Example 1.2.1 represents the way a typical calculus question might be asked, and Example 1.2.2 connects the same information to a particular application. All the knowledge needed for Example 1.2.1 is necessary to complete Example 1.2.2, but Example 1.2.2 also requires interpretation of the content from Example 1.2.1.

Note that the application problem in Example 1.2.2 is limited by the information given. Example 1.2.2, part 4 is deliberately asking a question about events not far into the future from those presented in the given graph. If one were estimating a movie's rating over 10 years, the information may be different related to the subject matter's falling in favor, cult status, etc.

Exercises

1. Consider a function $f(x)$ that has the following characteristic describing its end behavior:
$$\lim_{x \to \infty} f(x) = 1.$$

   (a) Sketch an example of $f(x)$ if $f(x)$ is decreasing on $(0,\infty)$.

   (b) Sketch an example of $f(x)$ if $f(x)$ is increasing on $(0, \infty)$.

   (c) Sketch an example of $f(x)$ if $f(x)$ cannot be characterized as either increasing or decreasing on the interval (a,∞) for any real number a.

2. Consider the function $f(x)$ presented in Figure 1.8. Describe aspects of the graph including intervals of increase or decrease, relative extrema, intervals of concavity, any inflection points, and end behavior. (You may want to use (i)-(x) within the text as an approach guide.) Estimate when necessary.

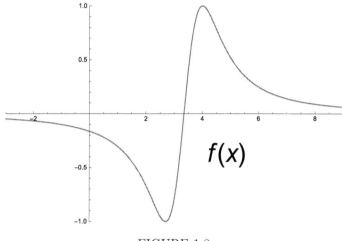

FIGURE 1.8

3. Consider the function $f(x)$ presented in Figure 1.9.

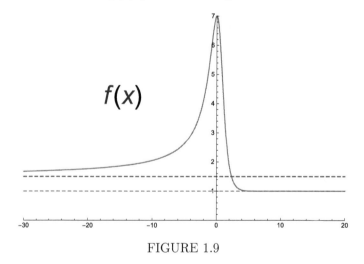

FIGURE 1.9

(a) Describe aspects of the graph including intervals of increase or decrease, relative extrema, intervals of concavity, any inflection points, and end behavior. (You may want to use (i)-(x) within the text as an approach guide.) Estimate when necessary.

(b) What is the maximum value of $f(x)$? In other words, what is the highest output value of $f(x)$?

4. Consider the function

$$P(t) = 30t^2 - \frac{16t^3}{3} + \frac{t^4}{4}$$

where P is the yearly profit (in thousands of dollars) of a toy producer t years after the first product is available for sale. A graph of $P(t)$ is shown in Figure 1.10.

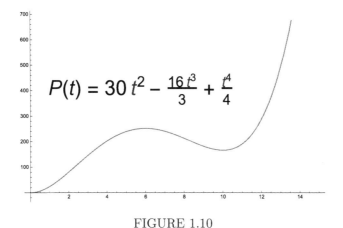

FIGURE 1.10

(a) Find the time when the yearly profit is highest in the first 8 years the company has been selling toys.

(b) When is the yearly profit decreasing?

(c) When is the yearly profit declining most rapidly?

5. Consider the function $B(x)$ presented in Figure 1.11. $B(x)$ describes the number of bacteria in a petri dish x minutes after bacteria are initially placed in the dish.

(a) Describe what is happening to the graph as $x \to \infty$. Then explain what that end behavior means about the number of bacteria in the petri dish.

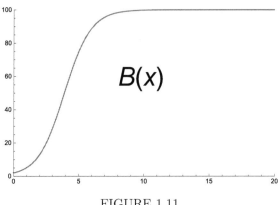

FIGURE 1.11

(b) Determine when $B(x)$ is increasing and when it is decreasing. Then explain what the intervals of increase and decrease indicate about the number of bacteria in the petri dish.

(c) Estimate when the number of bacteria is increasing the fastest.

6. The North American elephant populations are not self-sustaining owing to poor reproduction. Poor reproductive rates are largely due to the limited number of bulls in captivity. The zoo that you work for would like to bring in some bull (male) elephants to mate with their females; however, they ask you to determine the age of the bulls that they should bring in. You determine that the probability of fertility in bull elephants can be modeled by the function
$$p(t) = \frac{80000}{t^{\frac{3}{2}}} e^{-\left(\frac{30}{t} + t - 2\right)}$$
where t is the age of the elephant in years.

(a) Use your model for probability of fertility and calculus to make a recommendation as to the age of the bull elephant(s) that the zoo should bring in if the bull will only visit for a brief period of time.

(b) Use your model for probability of fertility and calculus to make a recommendation as to the age of the bull elephant(s) that the zoo should bring in if the bull will then remain at the zoo for a number of years.

(c) At what age might you tell the zoo that the elephant is too old to bring in to breed with their female elephants?

7. After they were protected from hunting, the elephants in Kruger National Park (South Africa) experienced birth rates exceeding death rates for 60 years. The researchers at Kruger National Park determined that the growth rate (birth rate-death rate) is at its maximum in 1983 where $t = 0$ in 1903. If we wish to model the elephant population from 1903 to 2013 with the logistic function,

$$f(t) = \frac{C}{1 + e^{\frac{k-t}{8}}},$$

pictured in Figure 1.12. determine C and k.

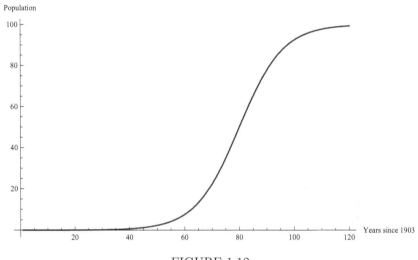

FIGURE 1.12

Project: Movie Ratings

Consider the following graph of $K(t)$ on $[0,8]$ (presented in Figure 1.13) as representing the average rating on a website (out of 5 stars) of the movie "Attack of the Killer Derivatives" t years after its release date. Assume viewers can continue to vote on the website throughout the 8 year period plotted. Answer the questions below and estimate values when necessary.

1. Immediately after the movie was released, what can you say about $K'(t)$? Did the rating get better or worse?

2. Determine all critical numbers of $K(t)$.

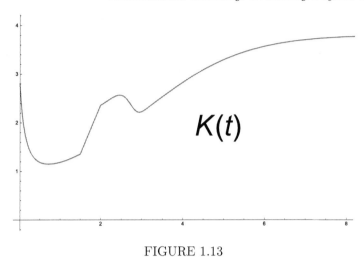

FIGURE 1.13

3. The movie's creator, a proud mathematician, was unhappy with ratings on the website. She decided to enlist the help of friends to vote on the website to artificially raise the average rating.

   (a) When did the mathematician and her friends intervene (related to the movie's release date)?

   (b) Explain, using calculus concepts, your answer to part 3a. What is happening at this point, points, interval, or intervals?

4. When the mathematician first began her quest to raise the rating, what was the trend of the rating? Do you think this is when she should have tried to affect the rating (assuming her idea was a good one)?

5. Does the average rating stabilize at a particular value? If so, what is that rating?

6. Let's say the mathematician and creator of "Attack of the Killer Derivatives" could choose a similar but more favorable graph for the ratings of the movie. What horizontal asymptote would be ideal for this graph? (Include the equation of the horizontal asymptote.) If she could have **any** function to represent the ratings, what would she pick as an explicit function for $K(t)$?

Project: Interpreting Snowball Melting Rates

Consider a spherical, melting snowball. The volume, V, of a sphere can be calculated

$$V = \frac{4}{3}\pi r^3 \tag{1.1}$$

where r is the radius of the sphere.

1. Consider V as a function of r. Then

$$V'(r) = 4\pi r^2. \tag{1.2}$$

   Plots of $V(r)$ and $V'(r)$ are presented in Figure 1.14. Determine which, of $V(r)$ and $V'(r)$, each graph (dashed or solid) represents and explain your choices.

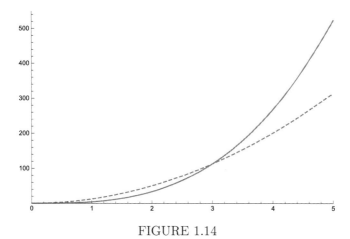

FIGURE 1.14

2. Based on your choice from part 1, where is $V(r)$ increasing and where is it decreasing? Explain how your answer(s) make sense with the volume of a sphere.

3. As the snowball melts, V and r can be considered decreasing functions of time, t. In order to connect the rates of change of V and r, the derivative can be taken with respect to t on both sides of (1.1)

$$\frac{dV}{dt} = 4\pi r^2 \frac{dr}{dt}. \tag{1.3}$$

   (a) If the snowball is melting, what can you say about $\frac{dr}{dt}$? What can you then say about the function $r(t)$?

    (b) If the snowball is melting, what can you say about $\frac{dV}{dt}$? What can you then say about the function $V(t)$?

    (c) Explain, in complete sentences, how your answers to parts 3a and 3b make sense with the equation in (1.3).

4. In parts 1 and 2, the volume, V, was considered as a function of r. In part 3, V and r were considered as connected functions of t. Explain how these different perspectives affect whether V is considered an increasing or decreasing function.

1.3   Transformations of Functions

Functions are often described *explicitly*, meaning that the output is calculated using a specific algebraic definition. Changes to the algebraic definition result in changes to output and, hence, to the graph. This section will focus on some ways of transforming functions and how to connect changes in function definition to graphical changes.

First, we will examine how to shift a graph up, down, right, or left. A function, $f(x)$, takes an element of its domain, x, and maps x to an element in the range, y. In order to shift a graph up, all output values should be increased by the same number. Conversely, to shift a graph down, all output values should be decreased by the same number. Hence, adding a number (positive or negative) to the entire function definition will move a graph up or down the y-axis.

In order to shift a graph right or left, the respective x value must change for each y value accordingly. In other words, the function should have the same output but change the input that goes with it by a certain amount. The horizontal axis or x-axis has smaller values to the left and larger values to the right. In order to shift to the left, we essentially want the outputs to occur earlier, so we could add a number to x before calculating the output. Similarly, we could subtract a number from x to make the outputs occur later, or shift the graph to the right. Hence, for a constant $k > 0$ and function $f(x)$:

- $f(x) + k$ will move $f(x)$ up k units,

- $f(x) - k$ will move $f(x)$ down k units,

- $f(x + k)$ will move $f(x)$ to the left k units, and

- $f(x - k)$ will move $f(x)$ to the right k units.

Multiplying by positive or negative numbers can change the shape of a graph. Recall for a differentiable function $f(x)$ and constant k that $\frac{d}{dx}[kf(x)] = kf'(x)$. If a function is multiplied by a constant, the derivative of that function will also be multiplied by a constant. Hence, the graph of a function can be made steeper or less steep by multiplying by a positive constant. Note that whether the graph is steeper or less steep will depend on whether k is greater than or less than 1. This type of transformation can also be referred to as a stretching or compressing of the graph.

When a function is multiplied by a negative constant, however, the outputs will be flipped over the x-axis (as well as affecting the steepness of the curve itself). Consider what happens when a function $f(x)$ is multiplied by -1. A point on the original curve of $(x, f(x))$ will be changed to $(x, -f(x))$ or, in other words, the sign of the y-coordinate will change. Hence, $-f(x)$ will be the reflection of $f(x)$ over the x-axis. If the sign of the x-coordinate is changed but the y-coordinate remains the same, the graph of $f(x)$ will be flipped over the y-axis. Hence, $f(-x)$ is a reflection of $f(x)$ over the y-axis. These transformations can be summarized for a constant $k > 0$ and function $f(x)$:

- $kf(x)$ will change the steepness of $f(x)$ by a factor of k,

- $-f(x)$ will reflect the graph of $f(x)$ over the x-axis, and

- $f(-x)$ will reflect the graph of $f(x)$ over the y-axis.

Note that not all possible graphical transformations are listed in this section. The transformations presented are intended as a beginning for understanding what models mean and how they can be changed. In order to understand any mathematical model, one must understand the structure of the equations or rules used.

Example 1.3.1 *Consider $f(x) = x^2$ and $g(x) = -4(x-1)^2 + 2$.*

1. *Explain how $g(x)$ can be graphed by making changes to the graph of $f(x)$.*

2. *Plot $f(x)$ and $g(x)$ on the same set of axes.*

1. When we consider how to change $f(x)$, we should work from the inside out. In other words, look at changes within the function first, then consider each piece calculated as if we were following orders of operation on $f(x)$. We can see that
$$g(x) = -4f(x-1) + 2.$$

The first transformation we should consider is that within the function, x is subtracted by 1. Hence we should take the graph of x^2 and shift it to the right 1 unit. After the calculation within the function, $f(x-1)$ is multiplied by -4. Multiplying the function by -4 will flip the graph over the x-axis and will stretch the graph to be steeper by a factor of 4. Lastly, the now downward-facing parabola will be shifted up 2 units because of the 2 added at the end of $g(x)$.

2. See Figure 1.15.

∎

When a graph or function is transformed, the shape of the graph may be changed considerably. In order to fully describe the way a graph has changed, we should also be able to articulate how the rates of change and extrema have changed.

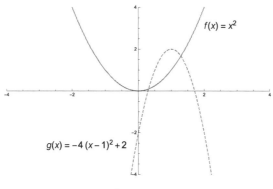

$f(x) = x^2$

$g(x) = -4\,(x-1)^2 + 2$

FIGURE 1.15

Example 1.3.2 *Consider $f(x) = x^2$ and $g(x) = -4(x-1)^2 + 2$ as presented visually in Figure 1.15.*

1. *State the types of any relative extrema (maxima or minima) and the locations where these extrema exist.*

2. *State the intervals of increase and decrease of $f(x)$.*

3. *Explain how the extrema (if any) of $f(x)$ have changed when $f(x)$ was transformed to $g(x)$.*

4. *Explain how the intervals of increase and decrease changed when $f(x)$ was transformed to $g(x)$.*

1. The function $f(x)$ has one relative minimum located at $x - 0$.

2. The function $f(x)$ is decreasing on $(-\infty,0)$ and increasing on $(0,\infty)$.

3. The function $g(x)$ still only has one relative extreme point (like $f(x)$), but instead of a minimum, $g(x)$ has one local maximum at $(1,2)$.

4. The function $g(x)$ still has only one interval of increase and one interval of decrease (as with $f(x)$). Because $f(x)$ was flipped over the x-axis in the transformation to $g(x)$, $g(x)$ begins by increasing on $(-\infty,1)$ and then decreases on $(1,\infty)$. The intervals of increase and decrease of $g(x)$ are also steeper than those of $f(x)$.

■

Exercises

1. Consider the function $f(x) = x^2$.

    (a) Create a new function $g(x) = f(x-2) + 1$. Write out an explicit function for $g(x)$.

    (b) Plot $f(x)$ and $g(x)$ on the same graph. Explain how the relative minimum of $f(x)$ is moved when the function is transformed into $g(x)$.

2. Consider the function $f(x) = |x|$. What does the function $f(-x)$ look like (relative to $f(x)$)? Explain.

3. Consider the two functions presented in Figure 1.16. The solid graph is $f(x) = \sqrt{x}$, and the dashed graph represents some function $g(x)$. Use the graph to answer the following questions about $g(x)$.

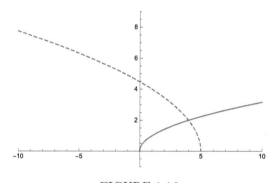

FIGURE 1.16

    (a) Briefly describe the visual differences in the two graphs. Then explain how $f(x)$ was transformed into $g(x)$. (Use complete sentences in your answer.)

    (b) Explain specifically how $g(x)$ is similar to and differs from $f(x)$ in terms of intervals of increase or decrease, relative extrema, intervals of concavity, inflection points, and end behavior.

    (c) Determine a possible explicit function $g(x)$.

4. Consider the two functions presented in Figure 1.17. The solid graph is $f(x) = x^3$, and the dashed graph represents some function $g(x)$. Use the graph to answer the following questions about $g(x)$.

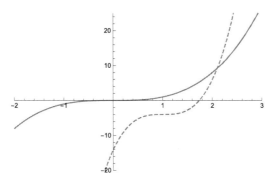

FIGURE 1.17

(a) Briefly describe the visual differences in the two graphs. Then explain how $f(x)$ was transformed into $g(x)$. (Use complete sentences in your answer.)

(b) Explain specifically how $g(x)$ is similar to and differs from $f(x)$ in terms of intervals of increase or decrease, relative extrema, intervals of concavity, inflection points, and end behavior.

(c) Determine a possible explicit function $g(x)$.

5. Consider the two functions presented in Figure 1.18. The solid graph is $f(x) = 4(x - 1)^2(x + 1)^2$, and the dashed graph represents some function $g(x)$. Use the graph to answer the following questions about $g(x)$.

(a) Briefly describe the visual differences in the two graphs. Then explain how $f(x)$ was transformed into $g(x)$. (Use complete sentences in your answer.)

(b) Explain specifically how $g(x)$ is similar to and differs from $f(x)$ in terms of intervals of increase or decrease, relative extrema, intervals of concavity, inflection points, and end behavior.

(c) Determine a possible explicit function $g(x)$.

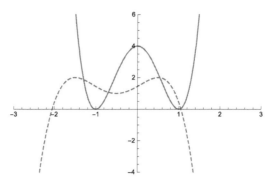

FIGURE 1.18

6. The North American elephant populations are not self-sustaining owing to poor reproduction. Poor reproductive rates are largely due to the limited number of bulls in captivity. The zoo that you work for would like to bring in some bull (male) elephants to mate with their females; however, they ask you to determine the age of the bulls that they should bring in. You determine that the probability of fertility in bull elephants can be modeled by the function

$$p(t) = \frac{80000}{t^{\frac{3}{2}}} e^{-\left(\frac{30}{t}+t-2\right)}$$

where t is the age of the elephant in years. Due to the limited number of bulls in captivity, zoologists have developed a drug to give to the elephants that shifts the development of their fertility to be one year later, but decreases the overall probability to 90% of what it is in elephants that are not on the drug.

(a) Determine a function $q(t)$ that describes the probability of fertility of bull elephants that take the new drug. Plot $p(t)$ and $q(t)$ on the same set of axes.

(b) The zoo you work for may have access to bull elephants that are on the new drug and those that are not. Make a recommendation concerning which type of bull elephant to choose to bring to the zoo based on the bull's age.

Project: Visual Transformation

Consider the two functions presented in Figure 1.19. The solid graph is a function $f(x)$, and the dashed graph represents another function $g(x)$. Use the graph to answer the following questions.

1. Briefly describe the visual differences in the two graphs. Then explain how

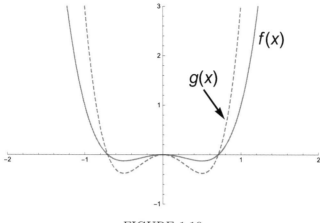

FIGURE 1.19

$f(x)$ was transformed into $g(x)$. (Use complete sentences in your answer.)

2. Sketch $f'(x)$ and $g'(x)$. Label your graph(s) well. Then explain what is different about the two sketches.

3. Sketch $f''(x)$ and $g''(x)$. Label your graph(s) well. Then explain what is different about the two sketches.

4. Summarize the differences in parts 2 and 3 and make conjectures on how your observations in part 1 led to these differences.

Project: Investigating an Explicit Transformation

Consider the two functions presented in Figure 1.20. The solid graph is $f(x) = 2x^3 - 3x^2 \quad 12x + 10$, and the dashed graph represents $g(x) = 5f(x) + 2$.

1. Write out an explicit formula for $g(x)$.

2. Find $f'(x)$. Solve an equation (either by hand or using technology) to determine the locations of all maxima and minima of $f(x)$ and to determine the intervals of increase and decrease.

3. Find $f''(x)$. Solve an equation (either by hand or using technology) to determine the locations of all inflection points of $f(x)$ and to determine the

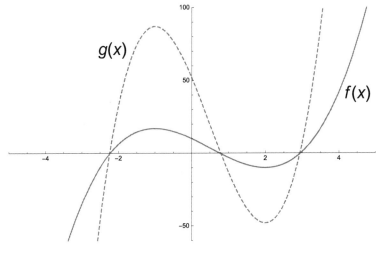

FIGURE 1.20

intervals of concavity.

4. Find $g'(x)$. Solve an equation (either by hand or using technology) to determine the locations of all maxima and minima of $g(x)$ and to determine the intervals of increase and decrease.

5. Find $g''(x)$. Solve an equation (either by hand or using technology) to determine the locations of all inflection points of $g(x)$ and to determine the intervals of concavity.

6. Summarize your findings in parts 2-5. Using the explicit formulas you found for first and second derivatives, explain how the transformation $5f(x) + 2$ changes the shape of $f(x)$.

1.4 Sensitivity

As shown in Section 1.3, changes to the explicit definition of a function can result in specific changes to a graph. When values used in a model change and result in significant changes in model output, we say that the model is *sensitive* to changes in those values. Understanding how changes to inputs or structure affect model output can help in understanding how a model is constructed and how to best improve a model.

Model sensitivity is important for several reasons. How sensitive a model is to a particular value may affect whether the model can be used to determine that particular value or if that value is important to the model itself. Investigating a model's sensitivity to particular fixed values, often called *parameters*, can also lead to a better understanding of how a model can be restructured or revised to more accurately predict the application being described.

Example 1.4.1 *Consider a vendor who has t-shirts for sale for $8 apiece. Let's say he has 15 t-shirts in total on this particular day, and the cost (to the vendor) of those t-shirts was a total of $50. His profit will then be his revenue from sales minus his cost or*

$$P(x) = 8x - 50$$

where $P(x)$ describes his profit in dollars as a function of x t-shirts sold. (Note that this profit function will only be accurate for integers $x \in [0,15]$.)

1. *Using a continuous $P(x)$, graphically show how the vendor's profit would change if his cost were $40 or $60.*

2. *Describe how changing the cost to the vendor affects the function $P(x)$.*

1. Representations of the original $P(x)$ (the profit function when the cost is $40) and the profit function when the cost is $60 are shown in Figure 1.21.

2. Intuitively, raising the vendor's cost should decrease his profit and the higher, dotted line in Figure 1.21 represents the vendor's profit when his cost is reduced from $50 to $40. In terms of the functions, the dashed line represents $y = 8x - 40$ or $y = P(x) + 10$. The graph of $P(x)$ is moved up the y-axis 10 units when representing the vendor's profit when his cost is $40. Similarly, the dashed line representing the situation when the vendor's cost is $60 is below the solid line representing $P(x)$, as we would expect because raising the vendor's cost should result in higher profit. The dotted line represents $y = 8x - 60$ or $y = P(x) - 10$, which means that $P(x)$ is moved down the x-axis 10 units. Changing the vendor's cost has a direct affect on the output; specifically the profit is increased by the number of

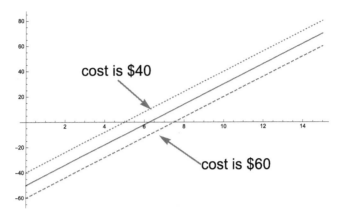

cost is $40

cost is $60

FIGURE 1.21

dollars the vendor's cost is reduced or is decreased by the number of dollars the vendor's cost is increased.

■

Note that the number needed to generate a profit is different for the various costs in Example 1.4.1. Originally, the vendor would need to sell 7 shirts to make a profit. If his cost is $40, he will need to sell 5 shirts to break even (and 6 to make a profit). If his cost is $60, he will need to sell at least 8 shirts to make a profit.

Example 1.4.2 *Again, consider a vendor who has t-shirts for sale as described in Example 1.4.1, but keep the vendor's cost fixed at $50.*

1. *Graphically show how the vendor's profit would change if his selling prices were $6 or $12 per shirt.*

2. *Describe how changing the vendor's selling price affects the function $P(x)$.*

1. Representations of the original $P(x)$ (the profit function when the price is $6 per shirt) and the profit function when the price is $12 per shirt are shown in Figure 1.22.

2. Intuitively, raising the vendor's selling price should raise his profit (as a function of items sold), and the dashed line representing when the price is $12/shirt in Figure 1.22 is above the solid line $P(x)$. Similarly, the dotted line representing when the price is $6/shirt is below the solid line $P(x)$. Raising the price makes the line steeper, and lowering the price results in a flatter line.

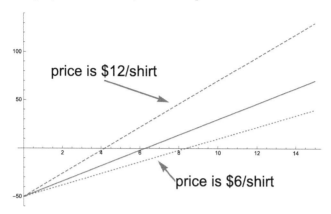

FIGURE 1.22

In Example 1.4.2, the slope of the line is being changed; hence, steeper and flatter lines based on those changes should be expected. Additionally, we can see that

$$
\begin{aligned}
y &= 12x - 50 = \frac{12}{8}(8x) - 50 = \frac{3}{2}(8x) - 50 + \frac{3}{2}50 - \frac{3}{2}50 \\
&= \frac{3}{2}(8x - 50) - 50 + 75 = \frac{3}{2}P(x) + 25
\end{aligned} \tag{1.4}
$$

which means that changing the slope from 8 to 12 stretches the line by a factor of $\frac{3}{2}$ and then moves the graph up the y-axis 25 units (or dollars). The line representing the situation when $m = 6$ can similarly be rewritten $y = \frac{3}{4}P(x) - \frac{25}{2}$.

However, restructuring models to clearly illustrate transformations, as shown in (1.4), can be difficult or even impossible. Being able to understand and articulate changes in model output may be more valuable. Note that by changing the sale price in Example 1.4.2, the number of t-shirts needed to make a profit also changes. When the t-shirt price is \$12/shirt, the vendor only needs to sell 4 t-shirts to make a profit and would need more than 8 to make a profit when the price is \$6/shirt.

Example 1.4.3 *Consider a function, $B(t)$,*

$$
B(t) = \frac{10}{50e^{-t} + 1}
$$

which describes the population of bunnies in hundreds in a particular forest as a function of time, t, in months after a particular starting time. A graph of this function is presented in Figure 1.23.

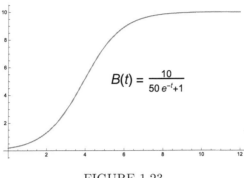

FIGURE 1.23

1. *One year after the starting time, has the population of bunnies settled to a particular value? What is the estimated population size at that time?*

2. *The settling value of $B = 10$ is sometimes referred to as the population's carrying capacity. Create multiple graphs varying this number in the function and comment on how changing the carrying capacity, or K, affects the graph of $B(t)$.*

3. *Now consider $B(t)$ with the numerator fixed at 10 but with various values where the number 50 is in the denominator. Create multiple graphs varying this number in the function and comment on how changing this number in the denominator affects the graph of $B(t)$.*

1. Looking at the graph, the curve appears to have a horizontal asymptote of $B = 10$, and the population is close to the asymptote after one year. We can also see that
$$\lim_{t \to \infty} \frac{10}{50e^{-t} + 1} = 10$$
Hence after one year, the bunnies are settling to an approximate population size of 1000 bunnies.

2. The carrying capacity could be lower or higher than 10, so $K = 5$, $K = 15$, and $K = 20$ were used to show differences in Figure 1.24. Changing the carrying capacity only slightly changes the shape of the overall curve. The rate of change is positive over the entire domain for all functions ($B(t)$ and the transformed functions), but the derivative itself is different for each curve. Each new carrying capacity can be thought of as 10 multiplied by

some number or $K = 10c$, and then the derivative of any of the new curves, $B_{new}(t)$, can be described

$$B(t) = \frac{10}{50e^{-t} + 1}$$

$$B_{new}(t) = \frac{K}{50e^{-t} + 1} = \frac{10c}{50e^{-t} + 1} = c\frac{10}{50e^{-t} + 1}$$

$$\Rightarrow B'_{new}(t) = cB'(t).$$

Hence, the new curves will be increasing more quickly or less quickly, depending on whether the carrying capacity has been increased (making the curve steeper) or the carrying capacity has been decreased (making the curve less steep).

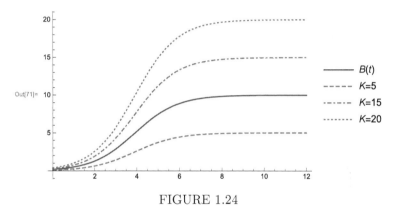

FIGURE 1.24

The curves with the new carrying capacities still exhibit leveling out end behavior, but the populations settle to different values (as we would expect).

3. Values for the varied coefficient were set to be 10, 25, 100, and 250, and the resulting plots are shown in Figure 1.25. As with the previous investigation, varying this number in the denominator only slightly changes the shape of the overall curve. The rate of change is positive over the entire domain, but the location of the inflection point has changed. In other words, changing this number affects where the bunny population is increasing the most rapidly. The smaller this number, the earlier the bunnies are increasing most rapidly and the population settles to the carrying capacity earlier.

∎

Note that all choices made in sensitivity investigations should be very clear. The graphs in Example 1.4.3 were labeled clearly to indicate what values and functions were presented. Additionally, values were chosen that created reasonable output. A negative carrying capacity would not make sense physically, and a negative coefficient in the denominator would create less appropriate plots.

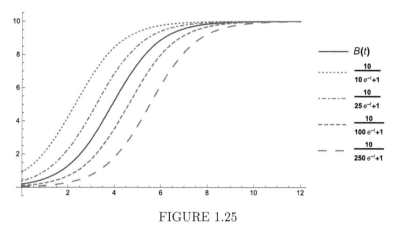

FIGURE 1.25

Exercises

1. Consider a linear function

$$f(x) = mx + b$$

   where m represents the slope of the line and b represents the y-intercept of the line. The constants m and b can be thought of as *parameters* of the line.

   (a) Changing which of the two parameters has a larger effect on $f'(x)$? Explain.

   (b) Changing which of the two parameters has a larger effect on the value of $f(0)$? Explain.

2. Consider an account that has an initial deposit of $2000 and earns a fixed annual interest rate of r, compounded continuously. The amount in the account after year t then can be determined by

$$A(t) = 2000e^{rt}.$$

   (a) Find $A'(t)$.

   (b) Create a graph with multiple plots representing the amount in the account when the interest rate is 1%, 2%, 3%, and 4% (or $r = 0.01$, $r = 0.02$, $r = 0.03$, and $r = 0.04$.)

   (c) Explain the effect changing r has on the output of $A(t)$.

3. Consider the function $A(t) = 100e^{-\frac{\ln(2)}{3}t}$ where A is the amount of a particular decaying chemical in grams after t years. The initial amount of the chemical present is 100 g and the half-life of the chemical is 3 years. (See Figure 1.1 in Example 1.1.1 for a graph of this function.) Plot $A(t)$ on the same graph with multiple other curves describing the amount of the same chemical remaining if the initial sample size is different. In other words, use other values than 100 g. After creating the graph, answer the following questions in full sentences.

   (a) How does changing the initial value size of the sample change the function's output values?

   (b) Does changing the initial sample size change the rate of change of the amount? If so, how?

   (c) What transformation has been performed on $A(t)$ for each secondary graph?

4. A ball is thrown straight up into the air. The height of the ball in meters (above the thrower's hand) is given by

$$h(t) = -t(9.8t - v_0)$$

   where t is time in seconds and v_0 is the initial velocity of the ball in m/s.

   (a) Verify that $h'(0) = v_0$.

   (b) Plot $h(t)$ for various values of v_0. Explain how changing the initial velocity of the ball changes the position of the ball before it returns to the thrower's hand.

   (c) Determine a relationship between v_0 and how long the ball is in the air.

5. The Penguin has opened his umbrella and is spinning it in front of Batman. The distance from the center to the end of one spike is 1 foot and the lowest spike is 2 feet above the floor. Batman throws a Batarang which lands at the end of one spike at the top of the umbrella (which is roughly depicted in Figure 1.26). The height of the Batarang from the floor can be described by

$$h(t) = \cos(at) + 3$$

   where h is in feet, t is in seconds, and a some fixed real number.

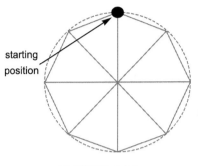

FIGURE 1.26

(a) Plot $h(t)$ for various values of a.

(b) What effect does the value of a have on the graph of $h(t)$?

(c) What, in the fight between the Penguin and Batman, affects the value of a?

6. After they were protected from hunting, the elephants in Kruger National Park (South Africa) experienced birth rates exceeding death rates for 60 years. The researchers at Kruger National Park determined the elephant population from 1903 to 2013 with

$$f(t) = \frac{C}{1 + e^{\frac{k-t}{8}}},$$

where $t = 0$ represents the beginning of the year 1903.

(a) Initially the researchers believed that the growth rate (birth rate-death rate) reached its maximum in 1983 (and that $k = 80$); however, they would like to know how this model is affected if the value of k is changed slightly. Fixing the value of $C = 100$, plot $f(t)$ for $k = 70, 75, 80, 85,$ and 90 and discuss how changing k affects the behavior of $f(t)$.

(b) The value C is the carrying capacity, the maximum number of elephants that the park can sustain; the park has attained a bit more land and would like to know what the model would look like if the carrying capacity is changed slightly. Fix the value of $k = 80$ and plot $f(t)$ for $C = 90, 95, 100, 105,$ and 110 and discuss how changing C affects the behavior of $f(t)$.

7. Zika, a mosquito born illness, is highly prevalent in tropical countries about 1 month after the rainy season due to issues with standing water and mosquito breeding. The mosquito population in a certain country can be modeled with the function

$$P(t) = 8000 \left(\frac{1}{6} + e^{-\frac{(t-c)^2}{d}} + e^{-\frac{(t-9)^2}{6}} \right)$$

shown in Figure 1.27.

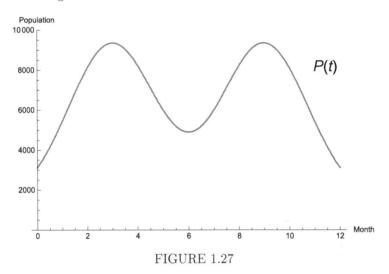

FIGURE 1.27

(a) In Figure 1.27, the first peak related to mosquito population occurs at month $t = 3$ (March), and the graph was plotted for $c = 3$. Plot multiple $P(t)$ curves using fixed $d = 6$ and $c = 1, 2, 3, 4$, and 5. Discuss how changing c affects the behavior of $P(t)$.

(b) Plot multiple $P(t)$ curves using fixed $c = 3$ and $d = 3, 6$, and 12. Discuss how changing d affects the behavior of $P(t)$.

8. (Sensitivity) Chris is trying to decide which pain reliever to take and is studying how long each drug stays in the body on average. The function $f(\tau) = e^{-k\tau}$ represents the percent of a certain drug in the bloodstream at time τ (in hours), where k is the rate of elimination of a single dose. (Note that $\tau = 0$ represents when the dosage is taken. Round all of your answers to two decimal places.)

(a) For acetaminophen, $k = .28$. Using this value for k, determine the percent of the drug in the bloodstream 12 hours after one dose is taken.

(b) If a single dose of acetaminophen is taken followed by another single dose 12 hours later, determine the amount of drug in the system 6 hours after the second dose is taken.

(c) For ibuprofen, $k = .295$. Using this value for k, determine the amount of the drug in the bloodstream 12 hours after one dose is taken.

(d) Based on your answers to parts 8a and 8c, how sensitive would you say the model is to small changes in the parameter k?

Project: Average Temperature

The graph in Figure 1.28 and the data in Table 1.1 represent average temperatures for each month in Greensboro, NC in a particular year [4]. The purpose of this project is to build a model, $T(t)$, to describe the average temperature T in °F as a function of month of the year, t. The overall average temperature is listed as 57.7°F.

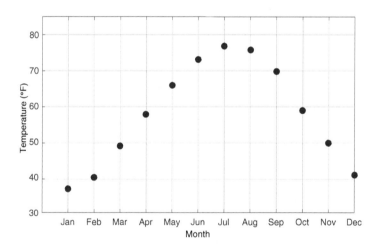

FIGURE 1.28

1. Begin by setting a number to each of the months for t. In other words, let January=1, February=2, etc. Plot the data.

2. The average temperature can be considered as part of a cycle, and hence, a trigonometric function may be appropriate to model the temperature.

TABLE 1.1

Month	Jan	Feb	Mar	Apr	May	Jun
Temp (°F)	36.7	39.9	48.7	57.6	65.8	73

Month	Jul	Aug	Sep	Oct	Nov	Dec
Temp (°F)	76.8	75.7	69.6	58.6	49.5	40.5

However, a basic sine or cosine curve will need to first be transformed to fit the data. The function $\cos(t)$ has a maximum at $t = 0$, but the maximum temperature does not occur at the beginning of the year. Find the number equivalent of the month with the highest average temperature. (We will later refer to this as t_{max}.)

3. The function $\cos(t)$ has a shorter wavelength than the graph in Figure 1.28. The normal period of $\cos(t)$ is 2π radians, but the period of the temperature model should be 12 months. Determine the factor, B, by which the period changes. In other words, find B such that $2\pi = 12B$.

4. The amplitude of the function $\cos(t)$ is 1. Find an amplitude, A, we can use for the model by finding the difference between the highest and lowest average temperatures and then dividing the difference by 2.

5. In order to model the average temperature, $\cos(t)$ needs to be moved up the y-axis. Set T_{avg} to be the overall temperature and then write the entire model

$$T(t) = A\cos(B(x - t_{max})) + T_{avg}.$$

6. Plot the model from part 5 with the data on the same graph.

7. The value of B is set by the length of a year, so the value of B is well known. We could make similar arguments about the values of A and T_{avg}. We set January=1, but we should probably interpret that value as indicating the beginning, middle, or end of the month. Regardless, the maximum may not be occurring exactly at a given data point. Investigate the sensitivity of the model to values of t_{max} by creating multiple curves for $T(t)$ (with different t_{max} values) on the same set of axes. Discuss (in full sentences) how the output graph is sensitive to changes in t_{max}. Also include justification for the values of t_{max} that you use (because you should only use values of t_{max}

that are fairly reasonable with the situation being modeled).

8. Based on your graphical investigation in part 7, choose a value for t_{max} that is reasonable and produces a nice fit to the data. (You should include graphs and/or calculations to justify your "nice fit.")

9. Recommend a model to fit the given data. Your recommendation should include your explicit function $T(t)$ with all units explained. Additionally, comment on any limitations or inaccuracies you see with the presented recommended function.

Project^{TE}: Nerve Activity

The purpose of this project is to try to use a particular sigmoidal equation to model the relationship between carotid distending pressures and sympathetic (nerve) activity. Software able to produce various graphs will be required for this project.

Consider the model

$$y(x) = \frac{1}{1 + (x/\alpha)^\beta}$$

where x represents the carotid distending pressure in mm Hg and $y(x)$ is muscle sympathetic activity (with normalized units, so it represents a percentage). The parameters α and β affect the shape and size of the curve $y(x)$. In order to find appropriate values for α and β, you will be investigating how each affects the graph of $y(x)$.

The data in Table 1.2 were taken and estimated from the right-hand side graph in Figure 1 of [2], which is a reproduction of Figure 4A in [3]. The data were then relatively normalized. (Note: the data were reduced in number and rounded considerably for the purposes of this project.) The input data (first value or x-value) is carotid distending pressure in mm Hg and the output data (second value, y-value) is muscle sympathetic activity (with normalized units). By changing α and β, you can investigate sensitivity, but it is your job to decide exactly what steps to take in your investigation.

Consider that you are a consultant who has been asked to find a particular model to describe this physiological system, and the product your employer wants is a report of your work. Answer the following questions in order to investigate the given model.

1. Plot the data and $y(x)$ on the same graph for $\alpha = 10$ and $\beta = 5$. Do these parameter values produce a curve that describes the data well?

2. Choose a particular value for α and keep that parameter fixed. Produce multiple graphs for $y(x)$ for various values of β with the same α value. What

**TABLE
1.2**

x	y
60	0.97
70	0.77
80	0.78
91	0.81
101	0.93
112	0.46
122	0.23
132	0.53
143	0.08
153	0.23
163	0.14
174	0.04
184	0.14
194	0.08

can you say about the sensitivity of $y(x)$ to changes in β (at this point)?

3. Choose a particular value for β and keep that parameter fixed. Produce multiple graphs for $y(x)$ for various values of α with the same β value. What can you say about the sensitivity of $y(x)$ to changes in α (at this point)?

4. Choose various values of α and β and create a systematic investigation of how these parameters affect the model $y(x)$.

5. Summarize your findings from part 4 using complete sentences and multiple paragraphs.

6. Based on your investigation, choose values for α and β that create a well-fitting (at least visually) graph to the given data. In a few complete sentences,

   (a) justify why you feel the graph is a good visual fit, and

   (b) explain how your work in parts 4-5 led to your specific choices for α and β.

Chapter Synthesis: Independent Investigation

1. (Revisit problem 5 from Section 1.1.) Consider the model

$$R(x) = \frac{25x}{x + 7}$$

   where R is the rate of metabolism of a particular enzyme in mg/h as a function of the concentration of the chemical being metabolized in mg/L.

   (a) Describe the behavior of $R(x)$ using techniques from Section 1.2.

   (b) In Section 1.1, a question asked for the maximum rate of metabolism of the enzyme. Using your work from part 1a, discuss the maximum rate of metabolism of the enzyme.

   (c) In Section 1.1, a question asked for the concentration of chemical present results in half the maximum metabolic rate. What is happening to the function at this point? Use your work from part 1a to discuss what is happening with the function.

2. ^{TE} (Revisit problem 5 from Section 1.3.) Consider the function

$$h(x) = m(x - c)^2(x + k)^2$$

   for constants m, c, and k.

   (a) Create plots as instructed below.
      i. Plot $h(x)$ for $m = 1$, $c = 0$, and $k = 0$.
      ii. Plot $h(x)$ for $m = 1$, $c = 1$, and $k = 1$.
      iii. Create a graph containing plots for $h(x)$ with $c = k = 1$ for various values of m.
      iv. Create a graph containing plots for $h(x)$ with $m = 1$ and $k = 1$ for various values of c.
      v. Create a graph containing plots for $h(x)$ with $m = 1$ and $c = 1$ for various values of k.
      vi. Create graphs containing plots for various values of all three constants.

   (b) Create a report that discusses the following:
      • the effect of changing m on the output, shape, and relative extrema of $h(x)$,

- the effect of changing c on the output, shape, and relative extrema of $h(x)$,
- the effect of changing k on the output, shape, and relative extrema of $h(x)$, and
- the effect of changing the three constants in combination (either in pairs or all three).

Include your graphs from part 2a within your report itself. Your report should include discussion of what you see in your graphs from part 2a as well as any additional graphs that help to support your investigation.

3. Section 1.1 contains a project on "Exponential Modeling," in which you are asked to find a function $P(t)$ that describes the number of bacteria present in a particular petri dish using an exponential model.

   (a) Formulate $P(t)$ as instructed in the original project, and then find $P'(t)$. Explain why $P'(t)$ is not reasonable to describe the rate of change in the number of bacteria for all time (because of limited resources or space in the petri dish).

   (b) Create a function called $G'(t)$ that better represents what you expect the *derivative* of the number of bacteria in the dish to be. (In other words, alter $P'(t)$ to create a different function that better models the rate of change of bacteria.) Explain why your new derivative is a better representation (and accounts for bacteria not being able to increase in number without bound).

   (c) If possible, find an explicit function $G(t)$ to better represent the number of bacteria in the petri dish. (Otherwise, explain what your $G'(t)$ indicates about $G(t)$ as fully as possible.) Include graphs of $G'(t)$, $G(t)$, $P'(t)$, and $P(t)$ and explain how $G(t)$ better represents the number of bacteria in the petri dish than $P(t)$.

4. (Revisit problem 6 from Section 1.3.) In more than one problem in this chapter, you have considered the role of a worker at a zoo who is trying to help select bull (male) elephants to bring in to mate with females. The North American elephant populations are not self-sustaining owing to poor reproduction, largely due to the limited number of bulls in captivity. You determined that the probability of fertility in bull elephants can be modeled by the function

$$p(t) = \frac{80000}{t^{\frac{3}{2}}} e^{-\left(\frac{30}{t} + t - 2\right)}$$

where t is the age of the elephant in years. Due to the limited number of bulls in captivity, zoologists want to develop a drug to give to the elephants that would help fight the low reproductive rates. You know from the drug developers that they have only been able to design drugs that either

- move the fertility window forward in time and lessen the probability of fertility (by a factor),

- lengthen the fertility window in time and lessen the probability of fertility (by a factor), or

- increase the probability of fertility but shorten the fertility window and lessen the probability of fertility (by a factor).

Create various functions that transform $p(t)$ in ways that fit the above criteria, then make a recommendation to zoologists of what drug would help the most in terms of type listed above as well as related quantities (such as percentages of fertility). Explain why the function and related drug would be helpful in increasing elephant reproduction numbers.

5. ^{TE} Section 1.4 contains a project on "Nerve Activity." Create a different function for the carotid distending pressure that could model the data. Rework the project using different values for constants in the function you develop.

2

Modeling with Calculus

Goals and Expectations

The following chapter is written toward students who have completed two or more semesters of collegiate calculus.

Goals:

- Section 2.1 (Using Derivatives and Rates of Change): To develop skills connecting derivatives to rates of change in context.

- Section 2.2 (Optimization):

    1. To review optimization concepts from introductory calculus.

    2. To use calculus to determine the best values in physical applications.

- Section 2.3 (Accumulation): To develop skills connecting integrals and the idea of accumulation to applications.

- Section 2.4 (Volumes): To use integrals calculating volumes of rotation in the context of applications.

- Section 2.5 (Sequences and Series): To use sequences and series to describe and learn about applications.

2.1   Using Derivatives and Rates of Change

The derivative of a differentiable function describes the original function's instantaneous rate of change at any particular input. When a differentiable function is used as a model to describe an application, the first and second derivatives of that function can provide valuable information about the situation being modeled. This section will focus on how derivatives can be used to learn more about applications.

Recall for a differentiable function $f(x)$:

- $f(x)$ is *increasing* when $f'(x) > 0$.

- $f(x)$ is *decreasing* when $f'(x) < 0$.

Hence, if a model is known to describe a particular situation, the derivative of that model could be used to learn about how quickly the output is increasing or decreasing. Further, the second derivative informs how the first derivative of a function is increasing or decreasing, and this acceleration of the function can also help us understand how an application is changing.

Example 2.1.1 *Patricia has $1000 dollars saved. She wants easy access to her money and does not want to risk losing any of it, so she decides to open a savings account at a local bank even though the interest rate will be very low. Patricia has identified options at different banks: (A) an account that has an annual interest rate of 0.5% compounded quarterly and (B) an account that has an annual interest rate of 0.45% compounded monthly.*

1. *Determine a function $A_1(t)$ that describes the amount in the account in dollars after t years if Patricia chooses the option (A). How much will be in her account after 5 years (to the nearest penny)?*

2. *Determine a function $A_2(t)$ that describes the amount in the account in dollars after t years if Patricia chooses the option (B). How much will be in her account after 5 years (to the nearest penny)?*

3. *Patricia does not know how long she will leave the money in the account. Use derivatives to explain which option is better in general.*

1. The function
$$A_1(t) = 1000 \left(1 + \frac{0.005}{4}\right)^{4t}$$

describes the amount in an account using option (A) in dollars after t years. Evaluating $A_1(5) \approx 1025.30$, Patricia would have \$1025.30 in an account using option (A) after 5 years.

2. The function
$$A_2(t) = 1000 \left(1 + \frac{0.0045}{12} \right)^{12t}$$

describes the amount in an account using option (B) in dollars after t years. Evaluating $A_2(5) \approx 1022.75$, Patricia would have \$1022.75 in an account using option (B) after 5 years.

3. Note that the general formula
$$A(t) = P \left(1 + \frac{r}{n} \right)^{nt}$$

can be differentiated using logarithmic differentiation

$$
\begin{aligned}
\ln(A(t)) &= \ln \left(P \left(1 + \frac{r}{n} \right)^{nt} \right) = \ln(P) + nt \ln \left(1 + \frac{r}{n} \right) \\
\frac{A'(t)}{A(t)} &= 0 + n \ln \left(1 + \frac{r}{n} \right) \\
\Rightarrow A'(t) &= n \ln \left(1 + \frac{r}{n} \right) \cdot A(t) = nP \left(1 + \frac{r}{n} \right)^{nt} \ln \left(1 + \frac{r}{n} \right)
\end{aligned}
$$

and the general derivative formula can be used to form the derivatives of A_1 and A_2:

$$A_1'(t) = 4000 \left(1 + \frac{0.005}{4} \right)^{4t} \ln \left(1 + \frac{0.005}{4} \right)$$

$$A_2'(t) = 12000 \left(1 + \frac{0.0045}{12} \right)^{12t} \ln \left(1 + \frac{0.0045}{12} \right)$$

Both functions $A_1'(t)$ and $A_2'(t)$ are strictly greater than zero, which means that $A_1(t)$ and $A_2(t)$ are both strictly increasing functions. The function with the higher rate (or higher derivative) will then be the better option (and would be the better option regardless of initial deposit). As shown in Figure 2.1, the function $A_1'(t) - A_2'(t)$ is positive, indicating that $A_1(t)$ is increasing at a higher rate than $A_2(t)$. Additionally, the difference in the rates of change appears to increase with time. Hence, option (A) is the better choice.

■

Note that a more in depth, by hand investigation could be used to investigate both $A_1'(t)$ and $A_2'(t)$. A briefer, graphical justification is used here in part 3 both for brevity and in order to focus on the argument and consider what information is contained in the derivatives.

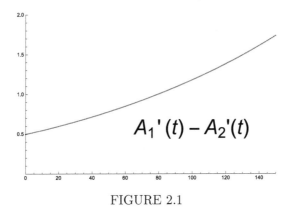

FIGURE 2.1

Exercises

1. Consider the function

$$f(x) = e^x$$

which has derivative $f'(x) = e^x$. Explain in your own words what $f(x) = f'(x)$ means about the shape of the original function.

2. Consider the function

$$P(t) = 30t^2 - \frac{16t^3}{3} + \frac{t^4}{4}$$

where P is the yearly profit (in thousands of dollars) of a toy producer t years after the first product is available for sale. Assume that $P(t)$ was based on data over a time period in the first 10 years, specifically $1 \le t \le 10$. The company wants to determine if it should continue making products after the end of year 10.

(a) Find $P'(t)$.

(b) Find $P'(10)$. This answer is how the model says yearly profit is changing at the end of year 10. Explain why this value *alone* does not indicate that the company should continue to make toys.

(c) Find $P'(11)$. Based on $P'(11)$, should the company continue to make toys? Why or why not?

3. Todd wants to analyze his spending and he has taken the data from his

credit card statements over one particular year and determined that the function

$$A(x) = 0.2423x^5 - 7.524x^4 + 82.814x^3 - 374.44x^2 + 551.15^x + 252.31$$

estimates his credit card balance in dollars as a function of x months since the beginning of the year. (For example, $x = 1$ would represent his balance at the end of January or at the beginning of February.)

(a) Plot $A(x)$.

(b) Find $A'(x)$.

(c) Determine when Todd's balance is increasing and when it is decreasing. In complete sentences, explain in words what times of year indicate a rising balance or declining balance.

(d) Determine when the balance is not only increasing, but the rate of increase is also increasing. In complete sentences, explain in words the time of year of your answer(s).

4. Tommy is planning on borrowing $10000. He will be charged a fixed annual interest rate of 5%, compounded quarterly. Tommy is planning on making a payment every quarter of $200. Data representing what he owes in dollars is plotted in Figure 2.2 with the x-axis representing years after the money was borrowed. Although the actual payments are discrete, the trend of his

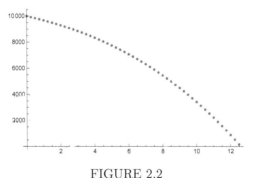

FIGURE 2.2

payments could be investigated as a continuous, decreasing function $A(t)$. Answer the following questions about a continuous function $A(t)$ that represents the trend of what Tommy owes in dollars as a function of time.

(a) Estimate when $A'(t)$ is increasing and when $A'(t)$ is decreasing.

(b) Use your answer to part 4a to explain when Tommy's owed balance is decreasing most rapidly.

5. Ben and Charles plan to run a cross country race against each other. Ben has observed Charles running this course previously and notes that Charles' distance function can be seen in Figure 2.3 where the y-axis represents miles run.

(a) Ben would like to run the race at the same speed throughout the entire race. At what speed would Ben have to run the race in order to guarantee that he wins the race (in miles/hour)?

(b) During which interval(s) of time is Charles running at a constant pace?

(c) If Ben and Charles are running side by side at the 10 minute mark and Charles' run from 10 to 12 minutes can be modeled by $\frac{4(x-11)^{1/3}+22}{16}$, when, in the interval 10 to 12 minutes, should Ben speed up to go as fast as he can for the rest of the race?

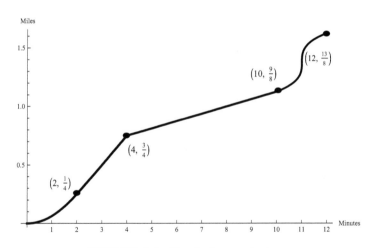

FIGURE 2.3: Charles' race strategy

Project: Savings When Paying Off a Mortgage

Molly has just purchased a house and wants to decrease the amount she pays in interest. She did a few calculations and was able to create data representing how much interest she would save (over the entire mortgage) if she paid $15 dollars more every month for the first few years. Molly likes having continuous functions to investigate, so she fit her data to a function and determined that

$$S(x) = 0.26x^3 - 20x^2 + 621x + 0.7$$

approximately determines the total money, S, in dollars saved (over the entire mortgage) if she overpays each month by $15 for x years. Answer the following questions related to the function $S(x)$ and how much money Molly can save with her plan.

1. Use $S(x)$ to approximate how much money Molly would save if she added $15 to each mortgage payment for the first year only. In other words, find $S(1)$.

2. Plot $S(x)$ and $S'(x)$ on separate graphs.

3. Find $S'(0)$, $S'(5)$, and $S'(10)$.

4. Home buyers are often given the advice of adding to the mortgage payments at the beginning of the mortgage because the money makes the most difference then. Use your previous work and your knowledge of derivatives to explain how the function $S(x)$ supports (or does not support) this suggestion.

5. Consider your plots of $S(x)$ and $S'(x)$ and the context of the function. For what interval of input values, x, should the function $S(x)$ be reasonable in the context of the application? Additionally discuss whether you think the function $S(x)$ could be reasonable on this entire interval.

Project^{TE}: Marginal Revenue, Cost, and Profit

Consider a particular manufacturer who is producing a new set of storage shelves. The manufacturer is considering selling the shelves (to retail outlets) for $20 apiece. The manufacturer has determined that they will have a fixed cost of $3000 for production of any shelves and additional costs of $g(x)$ dollars where

$$g(x) = 50\ln(x+1)$$

for x items produced, $x \geq 1$. Assuming all items produced are also sold, answer the following questions.

1. Revenue is the amount of money collected through sales. Determine a function $R(x)$ that defines the amount of money in dollars taken in by the manufacturer as a function of x items sold/produced.

2. Determine a function $C(x)$ that describes the cost in dollars of making x units.

3. Profit is defined by revenue minus cost. Form a function $P(x)$ that describes the profit of the manufacturer in dollars as a function of x items sold/produced.

4. Use technology to estimate how many shelves need to be sold to retail outlets for the manufacturer to make a profit.

5. Marginal revenue is the revenue generated by selling one additional unit and can be described by $R'(x)$. Find the marginal revenue value. Explain how your answer makes sense with the idea of marginal revenue.

6. Marginal cost is the cost of producing one additional unit and can be described by $C'(x)$. Find the marginal cost and plot this function. Explain what is happening to marginal cost as the number of items sold/produced increases.

7. Marginal profit is the difference between the marginal revenue and the marginal cost, or $P'(x)$. Find the marginal profit and plot this function. Explain what is happening to marginal profit as the number of items sold/produced increases.

8. How many shelves sold/produced result in the same value for marginal cost and marginal profit? What is marginal revenue at this x value? Explain why your answer makes sense with the relationship $P'(x) = R'(x) - C'(x)$.

Project: Investigating Snowball Melting Rates

Consider a spherical, melting snowball. The volume, V, of a sphere can be calculated

$$V = \frac{4}{3}\pi r^3 \tag{2.1}$$

where r is the radius of the sphere. As the snowball melts, V and r can be considered decreasing functions of time, t. In order to connect the rates of change of V and r, the derivative can be taken with respect to t on both sides of (2.1)

$$\frac{dV}{dt} = 4\pi r^2 \frac{dr}{dt}. \tag{2.2}$$

Equation (2.2) shows how the rates of change of $V(t)$ and $r(t)$ are connected, but the equation does not indicate how $V(t)$ or $r(t)$ change with time individually. Answer the following questions that investigate how given potential functions for $r(t)$ affect the related function $V(t)$.

1. The radius of the snowball is getting smaller with time; therefore, $r(t) = c$ for some constant does not make sense. Consider the following functions that potentially could model $r(t)$:

$$r_1(t) = -t + 5$$

$$r_2(t) = -\frac{1}{5}t + 5$$

where t is in minutes and r_1 and r_2 represent the radius of the snowball in centimeters.

(a) Plot $r_1(t)$ and $r_2(t)$ on a domain $[0,b]$ where b is the time when the snowball is completely melted.

(b) Find $r_1'(t)$.

(c) Find $r_2'(t)$.

(d) Plot $r_1'(t)$ and $r_2'(t)$ on a domain $[0,b]$ where b is the time when the snowball is completely melted.

(e) What is the initial radius of the snowball?

(f) When is the snowball completely melted?

(g) Explain in complete sentences how $r_1(t)$ and $r_2(t)$ are reasonable options to use to predict the radius of the melting snowball, and comment on their differences. (Include a summary of your results from parts 1e and 1f, but do not limit your arguments for the functions to a summary of these problems only.)

2. Use the following questions to begin investigating how $V(t)$ differs with the use of $r_1(t)$ and $r_2(t)$.

(a) Write out $V_1(t)$ and $V_2(t)$ incorporating the given formulas for $r_1(t)$ and $r_2(t)$ using

$$V = \frac{4}{3}\pi \left(r_1(t)\right)^3$$
$$V = \frac{4}{3}\pi \left(r_2(t)\right)^3 \ .$$

(b) Plot $V_1(t)$ and $V_2(t)$ on $[0,b]$.

3. Use the following questions to begin investigating how $V(t)$ differs with the use of $r_1(t)$ and $r_2(t)$.

(a) Use the following formulas based on (2.2)

$$\frac{dV_1}{dt} = 4\pi (r_1)^2 \frac{dr_1}{dt}$$
$$\frac{dV_2}{dt} = 4\pi (r_2)^2 \frac{dr_2}{dt}$$

the given formulas for $r_1(t)$ and $r_2(t)$, and the derivatives found in part 1 to write out derivatives for $V_1(t)$ and $V_2(t)$.

(b) Plot $V_1'(t)$ and $V_2'(t)$ on $[0,b]$.

4. Based on your work in parts 2 and 3, explain (in complete sentences) the similarities and differences of $V_1(t)$ and $V_2(t)$. Also include comments on which function for r you think is a better idea and why, or include an argument why you think both are essentially equivalent options. Also discuss at least one aspect of the physical problem that could be incorporated to help with the accuracy of the model.

2.2 Optimization

Most calculus I courses cover optimization in some way. Optimization itself means the process of optimizing a situation or creating the best outcome possible. One powerful tool in mathematical optimization is recognizing that a continuous function has relative high or low points when the derivative of that function changes sign. This section will focus on a review of calculus related to optimization as well as investigating situations where multiple aspects need to be considered to create the best mathematical outcome.

Recall the Extreme Value Theorem:

Theorem 1 (The Extreme Value Theorem) *If f is continuous on $[a,b]$, then f attains an absolute maximum value $f(x_1)$ and an absolute minimum value $f(x_2)$ for some numbers x_1 and x_2 in $[a,b]$.*

If the Extreme Value Theorem applies to a function $f(x)$ on $[a,b]$, there exists both a highest output and lowest output on the interval $[a,b]$. These absolute extrema can only occur at either

1. critical numbers (or where $f'(x) = 0$ or $f'(x)$ is undefined) or

2. endpoint a or b.

Hence, finding an optimal output for a continuous function is guaranteed, and the only possible locations for absolute extrema can be determined by considering the derivative and where the function is reasonable for the application. Additionally, behavior of the first and second derivative may help determine the location of absolute extrema of a function.

Example 2.2.1 *Consider the function $P(t) = 30t^2 - \frac{16t^3}{3} + \frac{t^4}{4}$ where P is the yearly profit (in thousands of dollars) of a toy producer t years after the first product is available for sale. At what instant is the business most profitable in the first 11 years after the first product is available for sale?*

In order to find the location of the maximum of the function, we first find the critical numbers.

$$\begin{aligned} P(t) &= 30t^2 - \frac{16t^3}{3} + \frac{t^4}{4} \\ P'(t) &= 60t - 16t^2 + t^3 \end{aligned} \tag{2.3}$$

The critical numbers of $P(t)$ are the values of t (in the domain of $P(t)$) that make $P'(t) = 0$ or make $P'(t)$ undefined. As shown in (2.3), $P'(t)$ is a polynomial so it is defined everywhere. Hence, the only critical numbers will occur when $P'(t) = 0$.

$$
\begin{aligned}
0 &= 60t - 16t^2 + t^3 \\
0 &= t(t^2 - 16t + 60t) \\
0 &= t(t - 6)(t - 10)
\end{aligned}
$$

The critical numbers of $P(t)$ are therefore 0, 6, and 10. We are only considering $P(t)$ on [0,11], and all critical numbers are in this interval. Because $P(t)$ is continuous on [0,11], we know that $P(t)$ obtains a maximum value on [0,11] (by the Extreme Value Theorem). Additionally, we know that the maximum value must occur at either a critical number (0, 6, or 10) or an endpoint (0 or 11). We can determine the maximum by evaluating $P(t)$ at each of these inputs:

$$
P(0) = 0
$$
$$
P(6) = 252
$$
$$
P(10) = \frac{500}{3} \approx 166.67
$$
$$
P(11) = \frac{2299}{12} \approx 191.58
$$

The highest output occurs at $t = 6$ which means that the business is most profitable at the end of the 6th year (or at the very beginning of the 7th year). ∎

Note that in Example 2.2.1, $t = 0$ was important because it was both a critical number and an endpoint. Also note that the example asked about a point in time; the maximum profit output value was approximately \$252,000.

Most optimization examples in basic calculus courses involve functions with a single critical number or with only one physically reasonable critical number. Functions based on applications may be more complicated, such as having more critical numbers (as in Example 2.2.1) or not being continuous. However, a strong grasp of function behavior can lead to greater flexibility in problem solving. The following list contains many concepts useful to determining an optimal situation described using mathematics:

- What are the critical numbers of the function?

- When is the derivative of the function positive or negative?

- When is the second derivative positive or negative?

- Does the function approach a horizontal asymptote as the input approaches infinity?

- What input values are reasonable with the given application?

The above list is not exhaustive, but the questions may be a good place to begin attempting real-world optimization problems.

Example 2.2.2 *A drug is administered into the body and the concentration of the drug, C (in mL/kg body mass), can be modeled by*

$$C(t) = 533(e^{-0.4t} - e^{-0.5t})$$

where t is the time in hours after the drug is administered. (Note that the equation above is not based on any particular drug, and the function itself and more information on pharmacokinetic modeling can be found in [7].) Determine when (after administration) the largest amount of the drug is in the body. Round the answer to the nearest hundredth hour.

$C(t)$ is a continuous function on any finite interval; however, we can see that

$$\lim_{t \to \infty} 533(e^{-0.4t} - e^{-0.5t}) = 0$$

and the only time that $C(t) = 0$ is when $t = 0$. Therefore, we really need to consider the interval $[0,\infty)$. Taking the derivative of $C(t)$

$$C'(t) = 533(-0.4e^{-0.4t} + 0.5e^{-0.5t})$$

we see that $C'(t)$ is defined everywhere, and hence the critical numbers of $C(t)$ can be found by setting $C'(t)$ equal to zero:

$$
\begin{aligned}
0 &= 533(-0.4e^{-0.4t} + 0.5e^{-0.5t}) \\
0 &= -0.4e^{-0.4t} + 0.5e^{-0.5t} \\
0.4e^{-0.4t} &= 0.5e^{-0.5t} \\
e^{-0.4t} &= \frac{5}{4}e^{-0.5t} \\
-0.4t &= \ln\left(\frac{5}{4}\right) + (-0.5t) \\
0.1t &= \ln\left(\frac{5}{4}\right) \\
t &= 10\ln\left(\frac{5}{4}\right) \approx 2.23 .
\end{aligned}
\tag{2.4}
$$

Because $C(t)$ is continuous on $[0,\infty)$ and has one (and only one) critical number, $C(t)$ will be either strictly increasing or strictly decreasing on $[0,10\ln\left(\frac{5}{4}\right))$. Further, $C(t)$ with be either strictly increasing or strictly decreasing on $\left(10\ln\left(\frac{5}{4}\right),\infty\right)$. Evaluation of any selected input in either interval reveals that $C'(t) > 0$ on $[0,10\ln\left(\frac{5}{4}\right))$ and $C'(t) < 0$ on $\left(10\ln\left(\frac{5}{4}\right),\infty\right)$. Hence, $C(t)$ increases in the interval before the critical number shown in Equation (2.4), and $C(t)$ decreases on the interval after the critical number. The drug is at its highest concentration in the body after about 2.23 hours. ∎

Exercises

1. A gardener wants to section off a rectangular garden into four pieces by laying dividing material in an X using the corners of the garden, as shown in Figure 2.4. If she has 30 feet of dividing material, what are the dimensions

FIGURE 2.4

of the largest overall rectangle she can create? Round each dimension to the nearest hundredth of a foot.

2. Todd wants to analyze his spending and he has taken the data from his credit card statements over one particular year and determined that the function

$$A(x) = 0.2423x^5 - 7.524x^4 + 82.814x^3 - 374.44x^2 + 551.15^x + 252.31$$

estimates his credit card balance in dollars as a function of x months since the beginning of the year. (For example, $x = 1$ would represent his balance at the end of January or at the beginning of February.)

(a) Use derivatives to determine the time when Todd's credit card balance is the highest in the one year period. In complete sentences, explain in words at what time of year this maximum occurs.

(b) Use derivatives to determine the time when Todd's credit card balance is the lowest in the one year period. In complete sentences, explain in words at what time of year this minimum occurs.

(c) Use derivatives to determine the time when Todd's credit card balance is increasing most rapidly. In complete sentences, explain in words at what time of year this incident occurs.

(d) Use derivatives to determine the time when Todd's credit card balance is decreasing most rapidly. In complete sentences, explain in words at what time of year this incident occurs.

3. A doctor would like to use an MRI scan to look at John Doe's clavicle (approximately the middle of the flat part of Mr. Doe's shoulder) on both the left and right sides of his body. The scan picks up some data and fits the curve

$$s(x) = (10 - 4x)^3(-3 + 2x)^3,$$

for x in $[1.3, 2.7]$, to the data.

(a) Assume that the location of John Doe's clavicle can be identified by determining where there is a horizontal tangent line to $s(x)$ but $s''(x) = 0$. Find the points at which John Doe's doctor would like to look at his clavicle, on each side of his body, based on the fit curve, $s(x)$.

(b) The top of John's head can be found where the absolute maximum occurs on the fit curve. Find the top of John's head using derivatives.

(c) The doctor plans on giving John injections at the trigger points on each side of his neck. These trigger points are located at the inflection points between the points found in part 3a and the top of John's head. Determine where the points of injection will be.

4. Farmer Jim has a chicken farm that starts out with 100 chickens. Unfortunately, farmer Jim has a population of 2 foxes that like to prey on his chickens. The rates of change of farmer Jim's chicken (c) farm population and the rate of change of the surrounding fox (f) population are directly related by the equations:

$$\frac{df}{dt} = -f + .01cf,$$
$$\frac{dc}{dt} = c - .01cf.$$

The behavior of chickens and fox populations in and around farmer Jim's farm can be seen in Figure 2.5.

(a) Based on Figure 2.5, what appears to be happening to the population of foxes when the chicken population reaches its maximum? Explain in context why this would be occurring.

(b) Based on Figure 2.5, what appears to be happening to the chicken population when the fox population reaches its maximum? Explain in context why this would be occurring.

(c) $c''(2.362671)$ is roughly 0. Explain what is happening to the chicken population when $t = 2.362671$ (think about what function is being

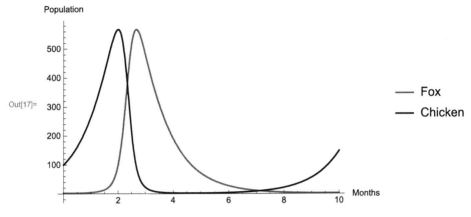

FIGURE 2.5: Fox versus Chicken Populations

optimized at this time).

(d) When $t = 2.362671$, $c(t) \approx 341$. Discuss what is happening to the chicken (or fox) population each time it reaches a size of 341.

5. ^{TE} George is determined to climb the peaks of Mathedor but is trying to find the best path to take. The peaks can be seen in Figure 2.6 and the height can be represented by the function

$$h(x,y) = \frac{2}{e^{(x-1)^2+(y-2)^2}} + \frac{1}{e^{(x+1)^2+y^2}} \, .$$

(a) George has decided to start his climb along the line $x = 0$. At what y value will George be between the two peaks? (Approximate your answer to 3 decimal places.)

(b) After George reaches his stop in part 5a, he decides to walk to the point $(1.272, 0.636, 0.292823)$ and follow along the line $x = \frac{y}{2}$ to climb to the top peak. Setting $x = \frac{y}{2}$ in $h(x,y)$, determine at which y value George will be working his hardest. (Approximate your answer to 3 decimal places.)

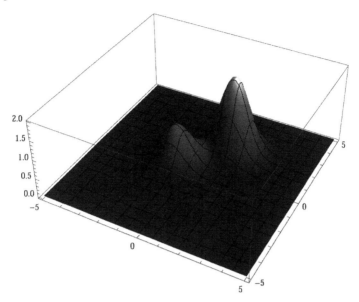

FIGURE 2.6: Peaks of Mathedor

Project: Calcu-cola vs. Pepsilon, Soda Price

Part A:

You are contacted by the company Calcu-cola to help them design cylindrical cans for their products. They want to optimally design the cans to hold 355 cm^3 of soda. (You may want to have an actual 12 oz soda can present.)

1. If the cans are perfect cylinders, what is the optimal radius for each can? Round your answer to 3 decimal places. (Hint: Minimize the surface area of the can.)

2. Is your answer to part 1 what you would expect for a typical soda can?

3. Soda cans, however, are not perfect cylinders. Explain, in a few sentences, differences in soda can construction that would change the optimized function in part 1. (You do not need to create any new functions. Just explain what possibly should be incorporated into the minimized function.)

4. One thing you should notice about actual soda cans is that the top and bottom of the can are thicker than the aluminum in the rest of the can. Rework part 1 letting the top and bottom of the can be twice as thick as the rest of the cylinder. Is your answer closer to what you would expect for

an actual soda can?

5. In full sentences, make a recommendation to Calcu-cola of the dimensions they should use for their cans (based on letting the bottom and top of the can be twice as thick as the rest of the can).

Part B:

Calcu-cola's new rival company, Pepsilon, is impressed by your work and is adopting dimensions of radius 3.1 cm and height 12.0 cm (with the top and base being twice as thick as the rest of the can) for their 12 oz soda cans. Because of your extensive knowledge on this matter, you have just been hired by Pepsilon to help them make additional decisions about their products.

The main question Pepsilon wants you to answer is this: what should they do to make as much money as possible on their new product, SineWave? In order to help you with this decision, the company gives you the following information:

- Pepsilon can buy aluminum to make the cans for 0.2 cents per square centimeter.

- It costs 1 cent per can to form each can and put soda in it.

- It costs 10 cents to make 100 fluid ounces of SineWave.

- Pepsilon spends $10,000 on marketing and all other costs.

You only have so much information, so you plan on using the same dimensions mentioned above for the can and you decide to assume that the number of cans sold is always the same as the number of cans produced.

1. Construct a function that describes the cost of the Pepsilon company to make SineWave. This function, $C(x)$, should be the cost as a function of x cans of SineWave produced (or sold). Since this is more than a function (it is also a model), when you write out the function equation also write out the units for the output, $C(x)$, and for the input, x. (Round the surface area of the can to the nearest square inch before forming $C(x)$.)

2. Explain what number of cans produced results in minimum cost.

Pepsilon does want to make money, not just spend it producing SineWave. As you have now learned, minimizing the cost does not really help you in this situation so you have to think about a more complicated situation than you did with your work at Calcu-cola. You know that a company's profit is the net money it takes in. In other words, total profit equals total revenue minus total cost or $P = R - C$. Revenue is all the money brought in through selling the product. The cost to produce the product is dependent on the number of items produced (variable costs) and some base amount of cost that the company spends regardless of how many items it produces (fixed costs).

You've already formed a cost function, $C(x)$, for the company; now, you want to think about describing the revenue. Pepsilon has not yet set a price for a can of SineWave. Use p to denote the price in dollars of selling one can of SineWave. Then the amount of money collected in revenue for SineWave can be described

$$R(x,p) = xp$$

where R is in dollars and is a function of **two** variables: x, the number of cans of SineWave produced, and p, the price in dollars of each can of SineWave.

Your job at Calcu-Cola only required you to know how to deal with functions of one variable, and Pepsilon didn't tell you there would be multivariable functions, but don't panic! Think about what you know and work from there.

3. You are using x to describe the number of items sold. Think about how the number of items sold relates to the price per item, $x = D(p)$. Discuss what $D(p)$ should look like. Do not use any equations here, but do use calculus terms (or possibly) notation to describe what $D(p)$ should look like.

Pepsilon did not have the forethought to do research on how much people will spend for a can of SineWave. So you have no data or information to go on when thinking about the function $D(p)$. This doesn't mean that you can't still amaze Pepsilon executives with your mathematical knowledge! In cases with limited or no data (and even in cases where you **do** have data), you should use what you know to build a model. You may not have data to do a statistical regression to find $D(p)$, but you should be able to reason out that $D(p)$ should be a decreasing function.

If you are modeling something and you do not know much about it, you should begin with simple assumptions about how things work. As price increases, we would expect demand to decrease. One way to model this would be to use an exponential function of some kind, like $D(p) = Ke^{-kp}$ for some positive constants K and k.

4. Let's say you want to start simply, so let's assume demand can be modeled by $D(p) = Ke^{-kp}$ for some positive constants K and k. (a) Form a function $P(p,K,k)$ to describe the profit in dollars as a function of price in dollars p and demand parameters K and k. (b) Under these assumptions and **REGARDLESS** of the value of K and k, what price must Pepsilon charge per can to make a profit?

5. Do you think the price you have found is reasonable? Why or why not?

6. List aspects of the problem that we did **NOT** incorporate in the profit function that could affect profit and/or suggestions on how we could have a more accurate profit function.

Project: Optimal Bees

In this project, we will think about how honeycombs are constructed and why. First, let's make some basic assumptions.

- The bees' honeycomb is made up of right prisms. In other words, the spaces in the honeycomb are three dimensional shapes with identical tops and bases and the height of each prism forms a right angle with the top and with the base.

- The tops and bases of the different right prisms are identical. In other words, the same prism is repeated over and over.

If a worker bee is constructing a honeycomb under the assumed conditions, what is the best shape for him to use? In order to answer this question, we'll investigate what we know about regular polygons (polygons with all sides of equal length) and about areas.

The area of a regular polygon can be calculated fairly easily. In order to illustrate the derivation of the formula, an example of an equilateral triangle is shown below in Figure 2.7. This particular equilateral triangle has the distance from any corner to the **circumcenter** (the point of intersection of the perpendicular bisectors) equal to 1. We could scale this shape so that this length is whatever we want, but for simplicity we will use 1. This distance is called the **circumradius**. The perpendicular distance from the circumcenter to any side is referred to as the **apothem** or **inradius** and is marked on the figure by the letter a. [6]

For an n-sided regular polygon, the area is $2n$ times the area of a triangle formed like the one in the diagram with height a. Further, all angles in a regular polygon have measure $\frac{\pi(n-2)}{n}$ radians and the circumradius bisects the angle of the polygon. Using trigonometry, you should see that

$$\sin\left(\frac{\pi(n-2)}{2n}\right) = \frac{a}{1} = a$$

and

$$\cos\left(\frac{\pi(n-2)}{2n}\right) = \frac{s/2}{1} = \frac{s}{2}.$$

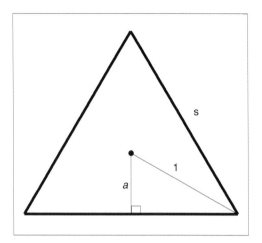

FIGURE 2.7

Since there are $2n$ of these right triangles within each regular polygon, the area of a regular polygon with circumradius 1 is

$$
\begin{aligned}
A &= 2n \cdot \frac{1}{2} \cdot \frac{s}{2} \cdot a \\
&= n \cdot \cos\left(\frac{\pi(n-2)}{2n}\right) \cdot \sin\left(\frac{\pi(n-2)}{2n}\right)
\end{aligned}
$$

Let's consider this area as a function of the number of sides, n. In order to make this easier to deal with, we'll use trigonometric identities to simplify:

$$
\begin{aligned}
A(n) &= n \cdot \cos\left(\frac{\pi(n-2)}{2n}\right) \cdot \sin\left(\frac{\pi(n-2)}{2n}\right) \\
&= n \cdot \frac{1}{2} \sin\left(\frac{\pi(n-2)}{n}\right) \\
&= \frac{n}{2} \sin\left(\pi - \frac{2\pi}{n}\right)
\end{aligned}
$$

Now that we know how to calculate the area of a regular polygon, we will consider what regular polygons of equal length circumradius have the greatest area.

1. Find $A'(n)$.

2. Graph $A'(n)$ and then explain what that tells you about what n leads to maximum area. (Remember to consider only reasonable values for n!)

3. Find $\lim\limits_{n \to \infty} A(n)$. Show all your work. (Show your steps.) Additionally, you should discuss how your result makes sense in terms of shapes.

4. If the bee wants to have the highest area for the top and base of the right prisms in the honeycomb and he wants to use a regular polygon (or something like it), what shape would you suggest he use?

Although maximizing the space used in each prism in the honeycomb is a good thing, the bee does not want to have any *wasted* space in the construction of his honeycomb. So he'd really like for all of the prisms to fit together without any space between them. If we consider a cross section of the collection of prisms, we would like them to **tessellate** the plane (or completely fill the two-dimensional space).

In order for a shape to tessellate the plane, the points of intersection of the adjacent polygons should have angles that add to 2π radians. Hence, an integer multiple of the angles must equal 2π for the shape to tessellate. Each regular polygon has internal angle $\frac{\pi(n-2)}{n}$.

5. Use the information above to identify all regular polygons that tessellate the plane.

6. Based on your answers to all of the questions in the project, what shape would you suggest the bee use as the top and base of his prisms? Justify your answer.

Project^{TE}: Calcu-cola vs. Pepsilon, IMPROVED Soda Price and Advertising Costs

Note:

- This project extends the work from the previous Calcu-Cola vs. Pepsilon project related to soda price (although this project can be completed without completing the original project).

You are working for the company Pepsilon and trying to come up with a price for each can of soda. Previously, you developed a model to describe the profit in dollars of Pepsilon but other workers at Pepsilon have managed to be more efficient and have reduced all general costs (except for marketing) from $10,000 to $1,000, so not incorporating marketing, you can now describe Pepsilon's profit by

$$P(p,K,k) = Ke^{-kp}(p - 0.61) - 1000$$

where profit (in dollars) depends on price in dollars, p, and demand function constants, K and k.

You have also been hard at work as a modeler and you have decided to update your model for profit, taking advertising into account. You believe that demand is directly related to the amount of money spent on advertising, so you revise your model by assuming that demand increases directly with increased advertising dollars or

$$K = ay$$

where y is the number of dollars spent on marketing and advertising and a is the multiplicative increase in demand per dollar (units: 1/dollar spent on advertising). For simplicity, we will assume that $k = 2.5$ and that a is some fixed parameter.

For this assignment, your overall goal is to make a recommendation to your boss at Pepsilon about appropriate values of p and y to maximize profit. You have no control over advertising directly, so you will need to consider multiple values for the parameter a. Your boss greatly believes in your abilities, but there are restrictions on resources. She cannot budget more than $10,000 in advertising.

1. Write out in (in full sentences) the profit function, P, you are using and what that function represents. Your function should have three inputs, p, a, and y, so that your function represents profit as a function of item price, dollars focused on advertising, and the parameter a.

2. Choose a few values for a. For each value, use mathematical software to plot the surface defined by $P(p,y)$. What affect does a have on P? Note: the output P will be the height at any point on the surface. (To plot a surface in *Mathematica* you can use the command "Plot3D"; to plot a surface in MATLAB®, you can use the command "surf.")

3. Explain what the parameter a represents in the application. What recommendation would you make concerning this parameter based on your answer to part 2?

4. Use mathematical software to find values of p and y that maximize the value of P. (If using *Mathematica*, you can use the command "FindMaximum"; to answer the question using MATLAB, you may want to use the "max" command or commands from the optimization toolbox.) Your answer should

make sense with the surfaces you plotted previously.

5. Summarize all your results and, in complete sentences, make recommendations to your boss concerning how to approach advertising and what price to set for the soda. Also comment on whether the price of the can of soda seems reasonable.

2.3   Accumulation

Integrals can be used to find the area below a curve. When the function being integrated models a rate of change, the area below this function can be considered an accumulation of that change. For example, distance is essentially the accumulation of velocity. This section will focus on how to view an integral as the accumulation of a particular quantity and how to connect the accumulation concept to applications.

Consider a car traveling directly away from a particular location with a velocity of exactly 60 mi/h. After 2 hours, the car will have traveled 120 miles. However, this same situation could be described by a velocity function $v(t) = 60$ where v is in mi/h and t is hours from a starting time. The area under the horizontal line $v(t)$ represents the accumulation of the velocity or the distance traveled, as shown in Figure 2.8. Note that the units of the area below the

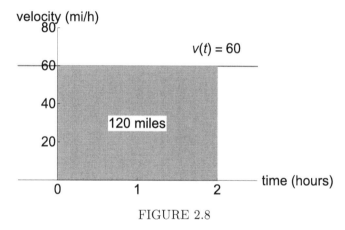

FIGURE 2.8

curve will be the product of the units of the input with the units of the output. The area under more complicated rate functions can also be calculated using integration.

Example 2.3.1 *Let the rate at which a population of bacteria changes in a petri dish (in bacteria/hour) be given by*

$$R(t) = \frac{4800e^t}{(24 + e^t)^2}$$

where t is the time in hours after the initial bacteria are placed in the petri dish.

1. *How many new bacteria grow in the petri dish in the second hour of the experiment? Round the answer to the nearest bacterium.*

2. *If there are 20 bacteria in the dish after 1 hour, about how many bacteria are in the dish after 2 hours?*

1. We can find the number of "accumulated" bacteria by evaluating the integral from $t = 1$ to $t = 2$:

$$\int_1^2 R(t)dt \quad = \quad \int_1^2 \frac{4800e^t}{(24 + e^t)^2}$$

$$\text{Let } u = 24 + e^t, \, du = e^t dt.$$

$$= \quad 4800 \int_{24+e^1}^{24+e^2} \frac{du}{u^2}$$

$$= \quad 4800 \left(\frac{1}{24 + e^1} - \frac{1}{24 + e^2} \right)$$

$$\approx \quad 26.73 \, .$$

Hence, the number of bacteria that grow in the second hour is about 27 bacteria.

2. Because 27 bacteria (approximately) grow in the dish in the second hour of the experiment, the total number of bacteria in the dish after 2 hours would be about 20+27=47 bacteria.

∎

Accumulation, however, is a measure of change, and the number of beginning bacteria was necessary to calculate how many total bacteria are in the dish in the previous example. Additionally, rates of change can be positive or negative. For example, Figure 2.8 represents when the car is moving away from a particular point and hence the car's distance from that point is increasing. The area below $v(t) = 60$ is above the x-axis and represents a positive accumulation. The number of bacteria in Example 2.3.1 is growing as well. More care would need to be taken when the velocity function is negative or changes sign.

Example 2.3.2 *A fish is swimming back and forth along one wall within its rectangular tank. The velocity of the fish can be described relative to a reference point just a little away from one wall perpendicular to the swimming route (see Figure 2.9). The rate of change in the fish's distance to the reference point (in inches per second) can be modeled by*

$$v(t) = 3 \sin \left(\frac{\pi t}{15} \right) .$$

If the fish is at the reference point at $t = 0$, what is the total distance the fish has traveled in the first 50 seconds (rounded to 2 decimal places)?

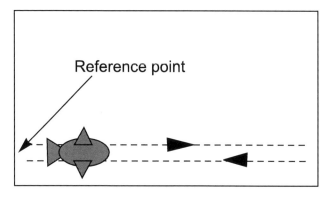

FIGURE 2.9: Fish swimming along one side of a tank.

The fish's velocity is zero whenever $t/15$ is a whole number. The fish is moving away from the reference point from $t = 0$ to $t = 15$, $t = 30$ to $t = 45$, $t = 60$ to $t = 75$, etc. and moving toward the reference point from $t = 15$ to $t = 30$, $t = 45$ to $t = 60$, etc. Four integrals can be constructed in order to find the total distance traveled by the fish, with the total distance traveled being the positive distances minus the negative distances:

$$
\begin{aligned}
\text{Distance} \quad &= \quad \int_0^{15} v(t)dt - \int_{15}^{30} v(t)dt + \int_{30}^{45} v(t)dt - \int_4^{50} 5v(t)dt \\
&\approx \quad 28.648 - (-28.648) + 28.648 - (-7.162) \\
&\approx \quad 93.106 \text{ inches}
\end{aligned}
$$

The fish traveled approximately 93.11 inches.

■

Exercises

1. A certain earthquake had disastrous effects. The graph of the probability of fatality versus the distance from the earthquake's epicenter can seen in Figure 2.10. Use Figure 2.10 to determine the probability of fatality between 4 and 9 km from the earthquake's epicenter.

2. Consider an account with beginning balance $500 that has interest compounded continuously at an annual rate of 1%. The amount of money in dollars in the account as a function of t years after the initial deposit is given by $A(t) = 500e^{0.01t}$.

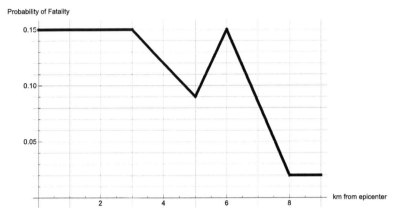

FIGURE 2.10: Probability of fatality versus distance (km) from epicenter.

(a) Find $A'(t)$.

(b) Find $\displaystyle\int_0^1 A'(t)dt$ to two decimal places. Explain what the value of this integral represents.

3. ^{TE} The probability that a pregnancy lasts exactly x days can be represented by the function $f(x) = \dfrac{1}{15\sqrt{2\pi}}e^{\frac{-(x-268)^2}{450}}$.

(a) Use $f(x)$ to determine the probability that a random woman's pregnancy would last less than 268 days.

(b) Use $f(x)$ to determine the chance that a random woman's pregnancy would last between 270 and 280 days. (Round to a whole percentage.)

4. ^{TE}We are going to consider modeling the administration of a drug to a particular organ. A person has taken a drug that needs to go to his liver. Assume that once the drug has been administered, the concentration of the drug in the patient's liver is represented by the function $c(t) = 2t^{3/2}e^{-t/2}$ mg/L. The total drug exposure over time is found by taking the area under the curve (from zero to infinity). Estimate the total amount of the drug the person had in his liver over the course of the treatment, as predicted by this model, to the nearest mg.

Project^{TE}: Is the Running App Working?

Eliza has decided to take up running and has installed an app on her phone. She is very interested in recording how fast she can run, but she's not sure the app is giving her accurate measurements of her speed. She runs for about 25 minutes where she knows the approximate distance she covers. The app on her phone records her pace 5 times, and those measurements are presented in Table 2.1. This project will use integrals to estimate Eliza's distance over the

TABLE 2.1

Time (minutes)	0	5	10	15	20
Pace (minutes/mile)	0	10	11	9	10

25 minutes. Answer the following questions.

1. Eliza knows the approximate distance she covered in miles. Convert the time data from Table 2.1 into hours, and convert the pace data to equivalent measurements in miles/hour.

2. The app also presents a function that is expected to approximately predict her speed in miles/hour as depending on time in hours:

$$f(x) = \frac{7e^{40x}}{10 + e^{40x}} - \frac{7}{10}.$$

Plot the converted data from part 1 and the function for her speed on the same graph.

3. Comment on how well (or how poorly) the function in part 2 predicts the data in Table 2.1.

4. Find an estimate of the distance she ran by evaluating $\int_0^b f(x)dx$ (where b is the time at the end of her run in hours).

5. Eliza knows the approximate distance she covers is 2.3 miles. Does the app appear to be giving accurate information?

Project: Credit Card Accumulation

Penny wanted to go to a concert and a night out with friends but did not have the cash to spend. She decided to put all her expenses on her credit card, which had no balance at the time. Her total expenses that night were $123.

Penny wants to pay off the debt quickly, so she decides she will pay $50 a month (or less) until the balance is zero. Her credit card calculates interest compounded continuously at an annual rate of 10.5%. (Note: actual credit card balances are usually compounded daily based on average balances. Continuous compounding is used in this project for simplicity.)

Penny is able to execute her plan of paying $50 (or less) each month. She pays $50 at the end of the first month, $50 at the end of the second month, and the remaining balance at the end of the third month. Her current balance on the account can be calculated by

$$A(t) = \begin{cases} 123e^{0.105t}, & 0 \leq t \leq \frac{1}{12}, \\ 73.4356e^{0.105t}, & \frac{1}{12} < t \leq \frac{1}{6}, \\ 24.303e^{0.105t}, & \frac{1}{6} < t \leq \frac{1}{4}. \end{cases}$$

1. Compounding continuously is modeled using the formula $A(t) = Pe^{rt}$ where P is the initial amount in the account in dollars, r is the fixed annual interest rate, t is the time after the initial amount is deposited in years, and A is the amount in the account in dollars at time t. The rate at which the account earns interest as a function of time, t, in years is $A'(t)$. Find $A'(t)$ for a general formula $A(t) = Pe^{rt}$.

2. Integrate the general $A'(t)$ from part 1 on a general interval $[a,b]$ and rewrite your answer to match the form below where you fill in the open box:

$$\text{Answer} = \boxed{} \int_a^b A(t)dt \ .$$

3. The integral of $A'(t)$ in part 1 represents the accumulation in an account compounded continuously. All three pieces of the piecewise function representing Penny's credit card balance fit the form of the formula in part 1. Use the formula from part 1 to set up three integrals that represent the accumulation in Penny's account over the three months she carries a balance. Then find the value of each integral.

4. Find the sum of the three integrals in part 3. This value should represent the interest Penny paid over the three months. Check your answer by considering what Penny paid in total for her $123 charges.

2.4   Volumes

Modeling three-dimensional situations may necessitate the use of multivariable functions, but calculations of some volumes can be calculated through integrals over one variable. Volumes of rotation can be used to describe some three-dimensional objects, and this section will focus on a brief review of volumes of rotation and their potential context. The focus of this section will not include integration techniques, and some integrals presented may be best estimated with mathematical software.

First, recall two methods for calculating volumes of rotation:

- The Disk (or Washer) Method and

- The Shell Method.

The Disk or Washer Method essentially stacks multiple thin coin-like layers, and the Shell Method puts multiple thin column "shells" within each other.

Consider rotating a rectangle around the x-axis, as shown in Figure 2.11. The rectangle has width Δx and height r, and when rotated around the x-axis, the height of the rectangle is the radius of the resulting cylinder. The volume of the produced cylinder would then be the product of π, the height, and the radius squared, or $\pi r^2 \Delta x$.

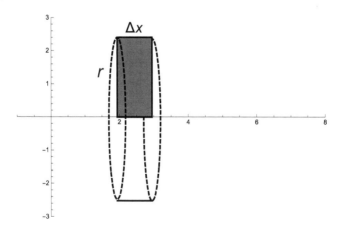

FIGURE 2.11

Multiple cylinders can be stacked with small thicknesses Δx. As the thickness of the cylinders decreases (and the number of cylinders increases), the overall stack will be closer to the volume of rotation. By taking the limit (as the number of cylinders goes to infinity) of the sum of each small disk volume, the volume of rotation of an area about a horizontal line $y = a$ can be found by

$$V = \pi \int_{x_0}^{x_1} ((r_1 - a)^2 - (r_0 - a)^2) dx$$

where the area being revolved is on the x interval $[x_0, x_1]$ and r_1 and r_0 represent the y-values of the top and bottom of the area, respectively. Example 2.4.1 goes through a volume of revolution problem using the Disk Method.

Example 2.4.1 *Find the volume generated by taking the area contained by* $f(x) = x$ *and* $g(x) = x^4$ *and rotating it around the line* $y = 2$.

The two functions intersect at two points, $(0,0)$ and $(1,1)$. A single disk is shown with the two functions in Figure 2.12, and the rotation about the horizontal line $y = 2$ will result in disks with holes in them (or "washers"). The

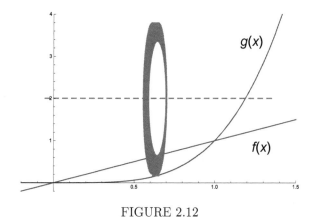

FIGURE 2.12

outer radius of each disk is calculated by $2 - g(x)$ and the inner radius of each disk is calculated by $2 - f(x)$. Hence, the volume generated can be found by evaluating the integral

$$V = \pi \int_0^1 ((2 - x^4)^2 - (2 - x)^2) dx .$$

This integral equals $\frac{44\pi}{45}$ or approximately 3.07.

■

Consider rotating a rectangle around the y-axis, as shown in Figure 2.13. The rectangle has width Δx and height h, and when rotated around the y-axis draws an open cylinder with radius r and thickness Δx. The volume of the produced cylinder could then be basically found by slicing and unrolling the "shell." The face of the unrolled cylinder has length and width based on the height of the cylinder, h, and the circumference of the circular base, $2\pi r^2$. Considering the resulting unrolled shape as a rectangular prism, the volume can be found by the product of the $2\pi r^2$ and the thickness Δx.

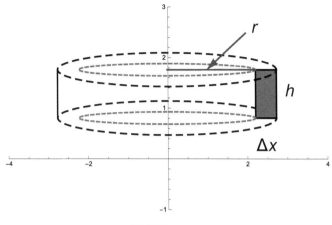

FIGURE 2.13

As with the Disk Method, the Shell Method uses a sum of multiple small volumes. As the thickness of the shells decreases (and the number of cylinders increases), the overall volume created by layering shells within each other will be closer to the volume of rotation. By taking the limit (as the number of cylinders goes to infinity) of the sum of each small shell volume, the volume of rotation of an area about the line $x = b$ can be found by

$$V = 2\pi \int_{x_0}^{x_1} (x - b)(h_1 - h_0)dx$$

where the area being revolved is on the x interval $[x_0, x_1]$ and h_1 and h_0 represent the y-values of the top and bottom of the area, respectively. Example 2.4.2 goes through a volume of revolution problem using the Shell Method.

Example 2.4.2 *Find the volume generated by taking the area contained by $f(x) = x$ and $g(x) = x^4$ and rotating it around the line $x = -\frac{1}{2}$.*

As mentioned in Example 2.4.1, $f(x)$ and $g(x)$ intersect at two points, $(0,0)$ and $(1,1)$. A single shell is shown with the two functions in Figure 2.14, and the rotation will result in shells with central axis $x = -\frac{1}{2}$. The shell radius will change as we move from the center layer to the outermost layer. Hence, we will integrate based on the interval of the area (from 0 to 1), and the radius of each shell will depend on progression through these x values. The radius of each shell can be described by the difference of the current location, x, and the x location

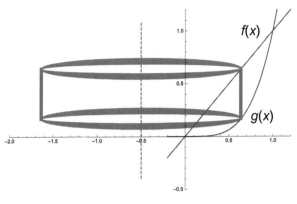

FIGURE 2.14

of the axis of location, and in this problem the radius of each shell is $x + \frac{1}{2}$. The height of each shell will be $f(x) - g(x)$, and hence the volume of revolution is calculated by evaluating the integral

$$V = 2\pi \int_0^1 \left(x + \frac{1}{2} \right) (x - x^4) dx .$$

The integral equals $\frac{19\pi}{30}$ or approximately 1.99.

■

Note that the calculated differences (such as to find a height or radius) are always determined by subtracting the larger number from the smaller number. In Example 2.4.1, "$1 - x$" was the calculation for the inner radius, and "$x - 1$" would have been incorrect. Additionally, the area in Example 2.4.2 uses "$x + \frac{1}{2}$" as opposed to "$-\frac{1}{2} - x$" because the x values over the integration (0 to 1) are larger than $-\frac{1}{2}$. See a traditional calculus textbook for additional information on the Disk/Washer and Shell Methods for finding volumes of rotation or for information on solving integrals.

Designing objects often involves restrictions on size or expectations on overall volume. Volumes of revolution can be used to investigate particular designs, as shown in the following example.

Example 2.4.3 *Ben works for a local ice cream shop that makes its own cones. Currently, they use a simple design, as shown in Figure 2.15. In the design a cylinder of diameter 2.25 inches and height 1 inch is on top of a truncated cone. The truncated cone base meets the cylinder with the same diameter, and the base of the overall ice cream cone has diameter 1 inch. The cylinder portion begins 2 inches above the bottom of the overall cone. Ben would like to improve on the style of the ice cream cone the shop uses. He would like to replace the*

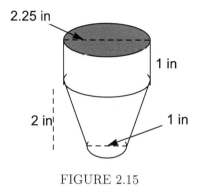

FIGURE 2.15

*truncated cone with a different shape. Ben generates an outer edge for the base
of the new cone by rotating the function*

$$f(x) = 2\sqrt[3]{x - \frac{1}{2}}$$

*about the y-axis for x values ranging from $x = \frac{1}{2}$ to $x = \frac{3}{2}$. The owner likes
Ben's idea, but he notices that the cylindrical portion on top would need to be
wider than before. The owner says he might consider the design, but Ben can
only use a cylinder with height $\frac{1}{2}$ inches. With this new stipulation, Ben's design
would look as shown in Figure 2.16. Will Ben's new design hold at least as much*

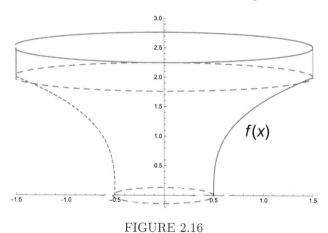

FIGURE 2.16

ice cream as the old design?

We can calculate the volume of the original cone using basic shapes. The cylinder at the top of the cone in Figure 2.15 is

$$V_{cyl} = \pi r^2 h = \pi \left(\frac{2.25}{2} \right)^2 (1) \approx 3.976$$

and the bottom of the cone can be calculated by subtracting the volume of the truncation

$$
\begin{aligned}
V_{bot} &= \frac{1}{3}\pi \left(\frac{2.25}{2} \right)^2 (2) - \frac{1}{3}\pi \left(\frac{1}{2} \right)^2 (2) \\
&= \frac{2}{3}\pi \left(\left(\frac{2.25}{2} \right)^2 - \left(\frac{1}{2} \right)^2 \right) \\
&\approx 2.127 \, .
\end{aligned}
$$

The overall volume of the original cone is then the volumes of the two pieces added together, which is about 6.103 in^3.

The volume of Ben's cone design can be found by (again) adding the cylindrical portion. The height of the top cylinder is $\frac{1}{2}$ inch, and the radius of the cylinder is $r = \frac{3}{2}$. Hence,

$$V_{cyl} = \pi \left(\frac{3}{2} \right)^2 \left(\frac{1}{2} \right) = \frac{9\pi}{8} \approx 3.534 \, .$$

The volume of the bottom portion of the cone can be found by rotating the region bounded by $f(x)$ and the base of the cylinder about the y-axis. We know that $f\left(\frac{3}{2}\right) = 2$; therefore the region being rotated is bounded on the right by $f(x)$, above by the line $y = 2$, on the left by the y-axis, and below by the x-axis. However, the function is only used for $\frac{1}{2} \le x \le \frac{3}{2}$. Hence we first should recognize that the center of the volume of rotation is a cylinder of volume $\pi \left(\frac{1}{2} \right)^2 2$ or $\frac{\pi}{2}$. The remaining volume is found by rotating the region bounded by $y = 2$ and $f(x)$ and $x = \frac{1}{2}$ about the y-axis. Figure 2.17 has the central cylinder shown in red and the rotated region outlined in black.

The volume of revolution can be calculated using the shell method

$$2\pi \int_0^{\frac{3}{2}} (x - 0)(2 - f(x))dx = 4\pi \int_0^{\frac{3}{2}} x \left(1 - \sqrt[3]{x - \frac{1}{2}} \right) dx = \frac{11\pi}{14}$$

and can be added to the central portion of the revolution, $\frac{\pi}{2}$, to find the total volume of the lower portion of the cone, $\frac{9\pi}{7}$. When both the top cylinder and the bottom portion of the cone are added together, the volume is $\frac{135\pi}{56} \approx 7.573$ in^3. Ben's new design would hold more ice cream than the old design.

■

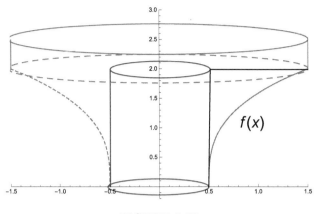

FIGURE 2.17

Exercises

1. A custom lighting shop is designing a new lampshade by rotating the area bounded by

$$f(x) = \frac{10 + 3x}{\sqrt{x - 2}} - 12$$
$$x = 0$$
$$y = 2$$
$$y = 7$$

about the y-axis. The function $f(x)$ and a schematic of how the area is rotated is presented in Figure 2.18. Note that x and y have units of inches. The lighting shop is interested in the total volume within the lampshade.

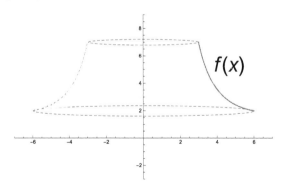

FIGURE 2.18

(a) Create an expression involving an integral that determines the volume of the lampshade.

(b) Find the total volume within the lampshade using the answer to part 1a.

2. ^{TE} Beth is creating a vase made out of pottery. She has planned out her design by rotating an area about the x-axis. The area rotated in her design is bounded by the functions

$$f(x) = \sqrt[4]{x}$$
$$g(x) = \sqrt[4]{x - 0.1} - 0.2$$
$$y = 0$$
$$x = 5$$

The functions $f(x)$ and $g(x)$ are shown in Figure 2.19. Note that x repre-

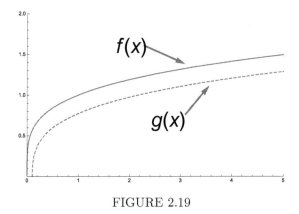

FIGURE 2.19

sents the distance in inches from the base of the vase and the output values of $f(x)$ and $g(x)$ represent the distance in inches from the center axis of the vase. Approximately how much clay (in cubic inches to 2 decimal places) does Beth need to complete her design?

3. After the work shown in Example 2.4.3, Ben's boss says that he cannot use Ben's new design because the volume is too large. Ben's suggested design from Example 2.4.3 would hold too much ice cream, and the owner of the ice cream shop does not want to increase his prices on cones. Ben changes his design so that the cone is entirely created by rotating the region bounded by

$$f(x) = 3\sqrt[3]{x - \frac{1}{2}},$$

the y-axis, the x-axis, and the line $y = 3$. Determine the volume of Ben's new design. Is the volume close to that of the current design (as presented in Figure 2.15 and Example 2.4.3)?

4. ^{TE} You are hired by the Grand Duke of Tuscany to make his bugles. You begin your exploration with smaller bugles but decide with some exploration to maximize the volume of the bugle.

   (a) Your clarinet shaped bugle is 24 inches in length with a bell diameter of 0.59 inches and can be modeled by rotating the curve $f(x) = \frac{1}{10x} + .195$ on the interval $[1,25]$ about the x-axis. Find the volume of your clarinet bugle (round to 3 decimal places).

   (b) Your trumpet shaped bugle is 58 inches in length with a bell diameter of 6 inches. Find a curve similar to $f(x)$ in part (a) that would satisfy your trumpet bugle's dimensions and then find the volume of your trumpet bugle.

   (c) Your tuba shaped bugle is 216 inches in length with a bell diameter of 17 inches. Find a curve similar to $f(x)$ in part (a) that would satisfy your tuba bugle's dimensions and then find the volume of your tuba bugle.

   (d) Finally, the Duke of Tuscany challenged you to make a bugle that uses an infinite amount of material and has a finite volume. What curve and interval will you choose to rotate about the x-axis in order to create this bugle?

5. Can We Float Inc. is charged with creating some new flotation devices, which will be attached to the arms, that will have at least 22 pounds of buoyancy.

   (a) For their first model, Can We Float decides to use the curve $f_1(x) = \sin(x)$ on the interval $\left[0, \frac{\pi}{6}\right]$ ft and rotate it about the x-axis. Find the volume of this first device.

(b) For their second model, Can We Float decides to use the curve $f_2(x) = x^2$ on the interval $[0,1]$ ft and rotate it about the x-axis. Find the volume of this second device.

(c) For their third model, Can We Float decides to use curve $f_3(x) = x^2$ on the interval $[0,1]$ ft and rotate it about the y-axis. Find the volume of this third device.

(d) The Can We Float devices use foam with a density of 1 kg/m^3. Assuming ordinary gravity and water with ordinary density, determine the buoyancy of each of the Can We Float devices in parts (a) through (c) using the formula B = volume of device in m^3 × density of material in kg/m^3 × 1000 × 9.81 $newtons/kg$.

(e) If regulation on a children's flotation device is 22 pounds of force and 1 newton is approximately 0.23 pounds of force, which of the Can We Float devices will satisfy regulation?

Project^{TE}: Promotional Footballs

A local sports store is producing small footballs as part of their promotion of a new store location. A graph of their proposed design for the footballs is below in Figure 2.20. The function $f(x)$ describes the distance from the central axis of the football to its edge in inches as a function of x inches from the end of the football. Answer the following related questions.

1. Set up an integral that calculates the volume of one football (in in^3).

2. Use a calculator or software program to calculate the volume of each football (in in^3).

3. Let's say the foam used for the footballs costs 8 cents/in^3 and the design and construction of each football costs 75 cents. The sports store wants to make 500 footballs to pass out at the next local football game. They estimate that this will increase their revenue by $1000. Should they go ahead with this promotion? Explain and show all your calculations. (Assume all costs are contained in the foam and design/construction costs.)

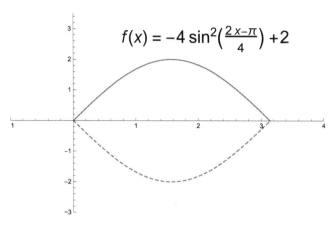

$$f(x) = -4\sin^2\left(\tfrac{2x-\pi}{4}\right) + 2$$

FIGURE 2.20

4. Assume that the cost for foam is still 8 cents/in^3 but that you can possibly negotiate a lower price for production because you have a large order. Determine a price for the construction per football that would make the promotion a good idea. Show all calculations and explain your recommendations (and reservations) in full sentences.

Project^{TE}: Calcu-cola vs. Pepsilon, Bottle Design

Pepsilon, the soda company, was very pleased with your previous work helping them understand price issues with cans of their new product SineWave. (See Projects in Section 2.2.) Now that SineWave is available to the masses, it has been flying off the shelves to mathematicians everywhere and Pepsilon wants to expand to sell the product in bottles. Pepsilon is requesting your services again in designing the bottles for SineWave.

1. In Figure 2.21, you will find a graph of a function $f(x)$ (the heavy solid line) that we will use to model the edge of a new bottle design. The function describing the outer edge of the bottle is

$$f(x) = 0.000369578x^6 - 0.0122x^5 + 0.1457x^4 - 0.76434x^3$$
$$+ 1.63174x^2 - 0.75127x + 0.6$$

where x is the distance in inches from the top of the bottle and $f(x)$ represents the distance in inches between the central axis of the bottle and the edge of the bottle. We need to know if this will be an appropriate design to hold the 500 mL of SineWave that Pepsilon wants to put in each bottle.

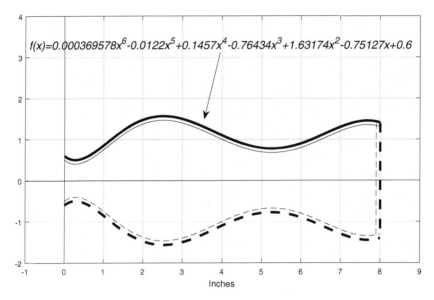

FIGURE 2.21

(a) Set up an integral to find the volume of revolution by rotating the area bounded by the function, the bottom of the bottle (at 8 inches), and the top of the liquid. Assume that the bottle is filled 1 inch from the top.

(b) Find the volume (in cubic inches).

(c) Convert the volume into mL, using the conversion: $1 \text{ in}^3 = 16.387 \text{ cm}^3$.

2. The liquid in the bottle will not go completely to the edge of the bottle, however. We need to recalculate the volume, accounting for the thickness of the plastic. On the graph in Figure 2.21, you will see a thinner curve below the given function.

(a) Set up an integral to find the volume of revolution representing the liquid in the bottle if the plastic is 0.1 inch thick. Again, assume that the bottle is filled 1 inch from the top.

(b) Find the volume (in cubic inches).

(c) Convert the volume into mL, using the conversion: 1 in^3=16.387 cm^3.

(d) How thick (in inches) would you suggest the bottle should be so that it contains 500 mL of SineWave?

3. (Sensitivity Analysis) The polynomial used to model the edge of the bottle has coefficients with several decimal places. Human error may lead to mistakes in these numbers. Let's investigate what happens when these coefficients are changed slightly. (For all answers below, use the given function and do not worry about the bottle thickness. Also, only change the function in the way directed in each specific problem.)

(a) Set up and calculate an integral to find the volume of revolution if the coefficient on x^6 is changed to 0.0004. Then convert this value to mL.

(b) Set up and calculate an integral to find the volume of revolution if the last digit is dropped from the coefficient on x^5. Then convert this value to mL.

(c) Set up and calculate an integral to find the volume of revolution if the last digit is dropped from the coefficient on x^4. Then convert this value to mL.

4. Summarize your results by writing (at least) a paragraph for each of the following.

(a) How thick (in inches) would you suggest the bottle should be so that it contains 500 mL of SineWave? Include a brief summary of your work leading to your suggestion.

(b) Describe pros and cons of the model we are using for the design of the SineWave bottle. (For example, what makes this a good model? What are problems with the model?)

(c) What other techniques or mathematics would you suggest using in the bottle design process? In other words, in what other way (or ways) could you model the structure of volume of the bottle?

2.5   Sequences and Series

In some cases, a discrete list or sum of a list better describes an application being modeled than a continuous function. This section will focus on a brief review of sequences and series and how they can be used in the context of modeling.

A *sequence* is a list of elements. For example, 1,2,3,4,5 is a finite sequence. Infinite sequences, such as

$$1, 2, 3, 4, 5, 6, 7, 8, \ldots \qquad (2.5)$$

have no end. Learning about the pattern of the sequence (if it has a pattern) can help in learning about the sequence itself. We say the sequence with nth element a_n *converges* if

$$\lim_{n \to \infty} a_n = L$$

and *diverges* if the limit does not exist (or is infinite). For example, the sequence presented in (2.5) continues to increase without bound, and hence, diverges. The sequence with elements $a_n = \frac{n+1}{n}$ for $n \geq 1$ or

$$2, \frac{3}{2}, \frac{4}{3}, \frac{5}{4}, \frac{6}{5}, \frac{7}{6}, \frac{8}{7}, \frac{9}{8}, \ldots \qquad (2.6)$$

converges because $a_n \to 1$ as $n \to \infty$.

Example 2.5.1 *Let an infinite list of coordinate points (x_n, y_n) be defined using the following sequences:*

$$x_n = \frac{7n + 90 \cos\left(30n + \frac{\pi}{9}\right)}{3n}$$

$$y_n = \frac{3\left(n + 10 \sin\left(30n + \frac{\pi}{9}\right)\right)}{n}$$

for $n \geq 1$. Are both sequences $\{x_n\}$ and $\{y_n\}$ convergent? If so, what is the location that the points are approaching?

Several points (x_n, y_n) for various values of n are plotted in Figure 2.22. In order to determine if the points are approaching a single location, the limit as $n \to \infty$

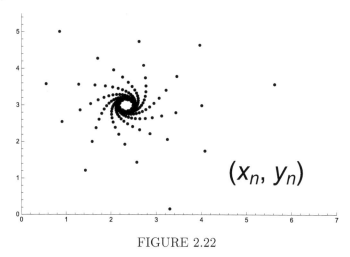

(x_n, y_n)

FIGURE 2.22

is taken of each sequence:

$$\lim_{n\to\infty} x_n = \lim_{n\to\infty} \frac{7n + 90\cos\left(30n + \frac{\pi}{9}\right)}{3n}$$

$$= \lim_{n\to\infty} \frac{7}{3} + 30 \lim_{n\to\infty} \frac{\cos\left(30n + \frac{\pi}{9}\right)}{3n}$$

$$= \frac{7}{3} + 30(0) \tag{2.7}$$

$$= \frac{7}{3}$$

$$\lim_{n\to\infty} y_n = \lim_{n\to\infty} \frac{3\left(n + 10\sin\left(30n + \frac{\pi}{9}\right)\right)}{n}$$

$$= \lim_{n\to\infty} 3 + 10 \lim_{n\to\infty} \frac{\sin\left(30n + \frac{\pi}{9}\right)}{n}$$

$$= 3 + 10(0) \tag{2.8}$$

$$= 3$$

Both sequences are convergent, and $(x_n, y_n) \to \left(\frac{7}{3}, 3\right)$ as $n \to \infty$.

■

Note that in steps (2.7) and (2.8) all steps were not shown. The limits that go to zero in both steps can be determined using the Squeeze Theorem, and those exercises are left to the reader. Also note that the center of the spiral in Figure 2.22 does appear to be $\left(\frac{7}{3}, 3\right)$, but determining if the points actually approach the center requires more than a visual examination.

Sometimes an application may concern the different values of a list, but in other situations a sum of all list values may be desired. Adding all the numbers in a finite list will always result in a number, but finding the sum of an infinite sequence is more challenging.

A *series* or *infinite series* is a sum of an infinite sequence. Series are typically represented using summation notation, such as

$$\sum_{n=n_0}^{\infty} a_n \tag{2.9}$$

where n_0 is the starting value of n (usually 0 or 1). A series *converges* or is *convergent* if this sum is finite. A series *diverges* or is *divergent* if the series is not convergent. While a series may be divergent because the sum becomes infinitely large, infinite sums are not the only way a series can be divergent. Some infinite sums simply do not approach any value. Consider the divergent series

$$\sum_{n=1}^{\infty} (-1)^n . \tag{2.10}$$

The series in (2.10) neither has a finite sum nor increases (or decreases) without bound. Each time a new element is added to the sum, the overall total switches from -1 to 0 or back.

Determination of the convergence or divergence of a series can be complicated, but the terms of a series must be getting smaller in magnitude as $n \to \infty$ in order for the overall sum to possibly approach a finite value. Hence, if the terms of a series do not approach zero, the series must be divergent. Recall the Divergence Test:

Theorem 2 (The Divergence Test) *Let a series be defined as presented in (2.9). If*

$$\lim_{n \to \infty} a_n \neq 0$$

then the series is divergent.

Note that the Divergence Test cannot be used to determine convergence of a series. A series may have terms that approach zero but still be divergent.

The *nth partial sum* of a series is defined by the sum of the first n terms of the series. If the terms of the series are defined a_n for $n \geq 1$, then the nth partial sum, S_n, is defined

$$S_n = \sum_{k=1}^{n} a_k$$

and a series is said to be convergent if

$$\lim_{n \to \infty} S_n = S$$

for some (finite) number S. The sum, S, of a convergent series is often difficult to determine. One type of series that the sum is direct to compute is a convergent *geometric series*.

A *geometric series* is a series of the form

$$\sum_{n=1}^{\infty} ar^{n-1} \tag{2.11}$$

where a and r are real numbers. Note that the geometric series above could be equivalently written

$$\sum_{n=0}^{\infty} ar^{n}. \tag{2.12}$$

As more fully described in standard calculus textbooks, the geometric series above in (2.11) or (2.12) is convergent if $|r| < 1$ and is divergent otherwise. If $|r| < 1$, the sum, S, of this geometric series can be calculated

$$\sum_{n=0}^{\infty} ar^{n} = \frac{a}{1-r}.$$

Example 2.5.2 *Jon has purchased a lottery ticket. If the ticket is a winner, Jon will receive $1000 immediately. Each year later for the rest of his life, Jon will receive half of the previous year's winnings.*

1. *What is the maximum amount that Jon could receive?*

2. *Essentially how much longer does the winner need to live to receive this amount?*

1. Jon's winnings can be described using a geometric series. If his ticket is the winner, he will receive $1000 immediately and $1000/2 in one year. These two payments are the first terms in the series corresponding to $n = 0$ and $n = 1$. If the winner lives forever, the winnings could be described

$$\sum_{n=0}^{\infty} 1000 \left(\frac{1}{2}\right)^{n} = \frac{1000}{1 - \frac{1}{2}} = 2000$$

and hence, the most Jon could receive would be $2000.

2. Trial and error can be used to show that

$$\sum_{n=0}^{17} 1000 \left(\frac{1}{2}\right)^{n} \approx 1999.99$$

$$\sum_{n=0}^{18} 1000 \left(\frac{1}{2}\right)^{n} \approx 2000.$$

Additionally, $S_n - S_{n-1} = ar^n = 1000 \cdot \left(\frac{1}{2}\right)^n$. In other words, the winner will be paid $1000 \cdot \left(\frac{1}{2}\right)^n$ in the nth year. Hence the payments in years 17, 18, and 19 will be

$$1000 \cdot \left(\frac{1}{2}\right)^{17} \approx 0.0076$$

$$1000 \cdot \left(\frac{1}{2}\right)^{18} \approx 0.0038$$

$$1000 \cdot \left(\frac{1}{2}\right)^{19} \approx 0.0019\,.$$

Between year 17 and year 18 the payment drops below a penny (with rounding). Hence, the winning ticket holder could be confident of collecting the maximum winnings by living 18 years (give or take a little, depending on how the payouts are rounded off).

■

Exercises

1. Let an infinite list of coordinate points (x_n, y_n) be defined using the following sequences

$$x_n = \frac{1}{n}\left[-5\cos\left(\frac{n}{3} - \frac{\pi}{6}\right) + 2n\sin\left(\frac{n}{20}\right)\right]$$
$$y_n = \frac{1}{n}\left[3n\cos\left(\frac{n}{20}\right) - 7\sin\left(\frac{n}{3} - \frac{\pi}{6}\right)\right]$$

for $n \geq 1$.

(a) Is the sequence $\{x_n\}$ convergent or divergent? Show all work.

(b) Is the sequence $\{y_n\}$ convergent or divergent? Show all work.

(c) Explain what is happening to the points as $n \to \infty$.

2. Let an infinite list of coordinate points (x_n, y_n) be defined using the following sequences

$$x_n = \frac{1}{2n}\left[3n + 10\cos\left(\frac{n}{2}\right)\right]$$
$$y_n = \frac{1}{n}\left[2n\cos(2n) + \sin\left(\frac{n}{2}\right)\right]$$

for $n \geq 1$.

(a) Is the sequence $\{x_n\}$ convergent or divergent? Show all work.

(b) Is the sequence $\{y_n\}$ convergent or divergent? Show all work.

(c) Explain what is happening to the points as $n \to \infty$.

3. A community in Sri Lanka currently has 3 pangolin inhabitants, $p_0 = 3$. This pangolin population grows using the following sequence

$$p_n = p_{n-1} + .45p_{n-1}\frac{80 - p_{n-1}}{80}.$$

(a) Determine the number of pangolin in this population when $n = 10$.

(b) If this pangolin population survives forever, will the population converge or diverge?

(c) Explain what is happening to the pangolin population as $n \to \infty$.

4. In a system there is a predator (s), and a prey (f). The predator and prey populations can be described by the sequences

$$s_{n+1} = .1s_n + .009s_nf_n,$$
$$f_{n+1} = 1.1f_n + .0005s_nf_n,$$

where $s_0 = f_0 = 100$.

(a) Find s_1 and f_1.

(b) If the predator and prey populations survive forever, will these populations converge or diverge?

5. Another community in Sri Lanka of 3 pangolin inhabitants grows using the following sequence

$$p_n = 3 + \sum_{k=0}^{n}\left(\frac{.45|80 - k|}{80}\right).$$

(a) Determine the number of pangolin in this population when $n = 10$ (round to the nearest pangolin).

(b) If this pangolin population survives forever, will the population converge or diverge?

6. Russell only has enough money to buy one lottery ticket between two options that cost the same amount and have essentially the same odds of winning. One ticket has a winning payout of $5000 immediately with a fifth of the previous year's winnings paid out each subsequent year. The other ticket will pay out $800 immediately and each subsequent year for 7 additional years (so 8 total payments).

   (a) If Russell expects to live forever, which should he choose?

   (b) How long does Russell need to live to make the choice from part 6a the better one?

7. It is the first day of class and Sam is wondering if anyone in the class will have the same birthday as him. Each student walks in one at a time and as they do Sam ponders the probability that no two people in the room will have the same birthday.

   (a) When the first person (other than Sam) walks into the room, what is the probability that Sam and this person share the same birthday?

   (b) When there are three people in the room, Sam and two others, what is the probability that none of them have the same birthday?

   (c) Use your answers from parts 7a and 7b to determine the probability that no two people in a group of $n < 365$ people share the same birthday.

   (d) We call Sam's problem the birthday problem. As more people enter the room, when would you be fairly certain that two people in the room share the same birthday?

   (e) As more people enter the room, what value does the birthday problem, the probability that two people in the room share the same birthday, converge to?

8. Sue has constructed a road for a car with square wheels so that the car travels on a smooth path (the center axis will stay at the same height) such as that seen in Figure 2.23.

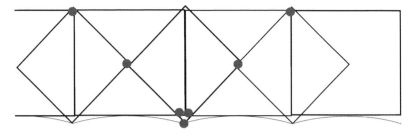

FIGURE 2.23: The path of a square wheel.

(a) Sue would like to follow the path of one point on the wheel throughout time. She has chosen to follow the height of the red point on the wheel shown in Figure 2.23. The height of the road is

$$h(x) = \begin{cases} -\cosh(x), & 0 < x < \sinh(1), \\ -\cosh(x - 2\sinh(1)), & \sinh(1) < x < 3\sinh(1), \\ -\cosh(x - 4\sinh(1)), & 3\sinh(1) < x < 5\sinh(1), \\ -\cosh(x - 6\sinh(1)), & 5\sinh(1) < x < 7\sinh(1), \end{cases}$$

where $\cosh(x) = \frac{e^x + e^{-x}}{2}$ and $\sinh(x) = \frac{e^x - e^{-x}}{2}$, and one of these arches is called a *catenary arch*. If the side length of the square is equal to the archlength of one catenary arch, determine the side length of the square.

(b) Determine the first 3 positions of the red point in Figure 2.23.

(c) If the square wheel continues on a similar path indefinitely, explain why the sequence defined by the height of the red point will never converge.

Project: A Series of Compounded Interest

When money is borrowed, interest is usually charged to the borrower and credited to the lender. Simple interest is calculated

$$I = Prt$$

where the amount A (owed or in an account) is calculated using the principal amount borrowed P, the interest rate r, and the time t. Banks more often calculate interest using some form of compounding, but calculations of compounding are based on the simple interest formula. Consider an annual interest rate r compounded over n periods per year in an account where P dollars has been deposited. When interest is compounded discretely, the interest within a single compounding period is calculated using the simple interest formula. Hence, the overall amount in the account at the end of a compounding period can be considered the terms of a sequence

$$A_0 = P$$

$$A_1 = P + Pr\left(\frac{1}{n}\right) = P + P\left(\frac{r}{n}\right) = P\left(1 + \frac{r}{n}\right)$$

$$A_2 = P\left(1 + \frac{r}{n}\right) + P\left(1 + \frac{r}{n}\right)\cdot r \cdot \left(\frac{1}{n}\right) = P\left(1 + \frac{r}{n}\right)^2$$

$$\vdots$$

$$A_k = P\left(1 + \frac{r}{n}\right)^k$$

which should resemble a familiar compounding function (when recognizing that k is the total number of compounding periods that have passed or nt). The total in the account (if held forever) could also be thought of as a series

$$A = P + \sum_{j=1}^{\infty} I_j$$

where I_j is the amount of interest charged or earned over compounding period j. The first few elements of the sequence defined A_k above are rewritten below to emphasize what interest is newly added in each step.

$$A_0 \quad = P$$

$$A_1 \quad = P + Pr\left(\frac{1}{n}\right) = P + P\left(\frac{r}{n}\right)$$

$$= A_0 + P\left(\frac{r}{n}\right)$$

$$A_2 \quad = P + P\left(\frac{r}{n}\right) + \left(P + P\left(\frac{r}{n}\right)\right)\cdot r \cdot \left(\frac{1}{n}\right) = P + P\left(\frac{r}{n}\right) + P\left(\frac{r}{n}\right) + P\left(\frac{r}{n}\right)^2$$

$$= A_1 + P\left(\frac{r}{n}\right) + P\left(\frac{r}{n}\right)^2 = A_1 + P\left(\frac{r}{n}\right)\left(1 + \frac{r}{n}\right)$$

$$A_3 \quad = P + P\left(\frac{r}{n}\right) + P\left(\frac{r}{n}\right) + P\left(\frac{r}{n}\right)^2 + \left[P + P\left(\frac{r}{n}\right) + P\left(\frac{r}{n}\right) + P\left(\frac{r}{n}\right)^2\right]\left(\frac{r}{n}\right)$$

$$= A_2 + P\left(\frac{r}{n}\right) + 2P\left(\frac{r}{n}\right)^2 + P\left(\frac{r}{n}\right)^3$$

$$= A_2 + P\left(\frac{r}{n}\right)\left(1 + \frac{r}{n}\right)^2$$

$$\vdots$$

$$A_k \quad = A_{k-1} + P\left(\frac{r}{n}\right)\left(1 + \frac{r}{n}\right)^{k-1}$$

In other words, we can describe the amount in the account using the sequence with elements A_k defined

$$A_k = P + \sum_{i=1}^{k} P\left(\frac{r}{n}\right)\left(1 + \frac{r}{n}\right)^{j-1}. \tag{2.13}$$

Use properties of sequences and series, (2.13), and other characteristics of compounded interest to answer the following questions.

1. Explain (for finite, positive P, r, and n) why the infinite sequence A_k ($k \geq 1$) is a divergent sequence.

2. Using properties of geometric series, explain why the series in the left hand side of (2.13) is divergent as $k \to \infty$.

3. Let \$500 be borrowed at 10% interest compounded monthly. Write out Equation (2.13) to describe the amount owed after k months (assuming no money has been repaid).

4. Assume the same situation as described in part 3. You have borrowed the money and are working out a way to pay off the debt. You would like to pay the debt off in 10 months. If the debt is paid in full at the end of 10 months, how much will you pay? (In other words, calculate the amount of debt after 10 months.)

5. Assume the same situation as described in part 3, and you want to pay off the debt in 10 months, but you want to spread out the payments over the 10 months. Set up (and write using summation notation) an equation describing the amount owed A_k if you pay \$10 at the end of one month, \$20 at the end of two months, and \$10k at the end of the kth month until the last payment is less than \$10k. (You may be general in your equation and assume it describes an amount prior to the final payment.)

6. Using the payment plan described in part 5, determine the amount of your last payment. Also determine the total amount repaid to the bank.

7. How much money was saved by using the plan in part 5 over the single payment plan described in part 4?

Chapter Synthesis: Independent Investigation

1. ^{TE} (Revisit problem 4 from Section 2.3.) We are going to consider modeling the administration of a drug to a particular organ. A person has taken a drug that needs to go to his liver. Assume that once the drug has been administered, the concentration of the drug in the patient's liver is represented by the function $c(t) = at^{3/2}e^{-t/2}$ mg/L for some constant a. The total drug exposure over time is found by taking the area under the curve (from zero to infinity).

   (a) (Revisiting the original problem) Estimate the total amount of the drug the person had in his liver over the course of the treatment as predicted by this model, if $a = 2$, to the nearest mg.

   (b) (Revisiting Section 1.3) Based on what you know about transforming functions, how do you expect changing the value of a will change the shape of the curve $c(t)$?

   (c) (Revisiting Section 1.4) Using various values of a (including 2, values less than 2, and values greater than 2), plot $c(t)$ and discuss how changes in a change the shape of the curve $c(t)$.

   (d) (Revisiting Section 1.4) Using various values of a (including 2, values less than 2, and values greater than 2), investigate how changes in a affect the total amount of the drug in the liver over the course of treatment.

   (e) The company that makes the drug knows that a concentration at any time of 7 mg/L would be detrimental, but they want the total amount during the course of treatment to be as close to 40 mg as possible. What would a need to be for the patient to receive the appropriate dose but not reach inappropriate levels?

2. Section 2.5 contains a project on "A Series of Compounded Interest." Assume, as in the project, that you borrowed $500 at 10% interest, and that your finances limit you to the following restrictions:

   • You have at most $40 to put toward the debt every three months.

   • You can pay no more than $20 toward the debt in any given month, but if you pay $15 dollars or more in one month, the following month you can only afford a $5 payment.

   • The bank that loaned the money requires at least $5 each month.

Design a payment plan that works under the above conditions but minimizes your interest payment.

3. ^{TE} (Revisit Example 2.4.3 and problem 3 from Section 2.4.) The owner of the ice cream shop likes Ben's enthusiasm for design and would like to change from his current cone (as shown in Figure 2.15). The owner allows any worker in the shop to try to create a cone that follows the following guidelines:

   - The cone holds at least 5.5 in^3 of ice cream but no more than 7 in^3 of ice cream.

   - The cone is easy to hold.

   - The amount of material needed for construction of the cone will not be very different from the current design.

   As a worker in the shop, create a new design for the cone that is in line with what the owner has requested. Write a report that includes details of your new design and how the design would be beneficial to the owner.

4. ^{TE} Section 2.4 contatins a project on "Calcu-cola vs. Pepsilon, Bottle Design." Develop a new function $f(x)$ to model the outer edge of the bottle. Again, account for thickness for the bottle material and verify that your design holds the expected amount of liquid. Create a report arguing why your design is a more suitable choice than the one presented in the project.

5. Section 2.1 contains a project on "Investigating Snowball Melting Rates." Using a similar approach, investigate the melting rates of different shaped ice "cubes."

6. Section 2.1 contains a project on "Investigating Snowball Melting Rates." Imagine you can develop a synthetic snow for snowballs that melts a particular way. Create a design for the melting properties of your snow by defining functions for how the surface area of a spherical snowball changes, $\frac{dA}{dt}$. Investigate how the volume of the snowball would change as a result, and justify which of your functions would result in a better design for more slowly melting snowballs.

3

Modeling with Linear Algebra

Goals and Expectations

The following chapter is written toward students who have completed a course in collegiate linear algebra.

Goals:

- Section 3.1 (Systems of Linear Equations): To introduce the student to a variety of applications related to topics in linear algebra, including systems of linear equations, eigenvalues and eigenvectors, rowspace and nullspace.

- Section 3.2 (Discrete Stochastic Processes):

    1. To review the concept of Markov chains.
    2. To introduce the student to real world applications related to discrete stochastic processes.
    3. To introduce the idea of model sensitivity to parameters in relation to linear models.

- Section 3.3 (Optimization):

    1. To introduce the student to the terminology related to constraint based optimization.
    2. To investigate different techniques for solving constraint based optimization problems.
    3. To introduce the student to the simplex method algorithm.

- Section 3.4 (Linear Techniques for Data Analysis-Least Squares Regression):

    1. To review the concept of least square regression.
    2. To expose the student to a linear algebraic derivation of linear regression.
    3. To investigate a variety of applications of linear regression.
    4. To introduce the student to basic concepts of error analysis.

- Section 3.5 (Linear Techniques for Data Analysis-Seriation):

  1. To introduce the student to basic cost functions used with ordering functions.

  2. To introduce the student to the concept of seriation and its applications.

- Section 3.6 (Linear Techniques for Data Analysis-Singular Value Decomposition): To introduce the student to the concept of singular value decomposition and how it can be used in modeling.

3.1 Systems of Linear Equations

As we learned in linear algebra, a system of linear equations can have one solution, no solution, or infinitely many solutions. Recall that a linear system $Ax = b$ has a unique solution for every vector b if and only if the matrix A is invertible.

In modeling, there are problems in which it is important to identify a unique solution and then there are other problems where it is equally important to just identify that a solution exists. So let's start with an example in which the goal of the problem is to find a unique solution.

Example 3.1.1 *Equity Autos manufactures luxury cars and trucks. The company believes that its most likely customers are high-income women and men. To reach these groups, Equity Autos has embarked on an ambitious TV advertising campaign and has decided to purchase 1-minute commercial spots on two types of programs: comedy shows and football games. Each comedy commercial is seen by 7 million high-income women and 2 million high-income men. Each football commercial is seen by 2 million high-income women and 12 million high-income men. Equity would like the commercials to be seen by 40 million high-income women and 80 million high-income men.*

If f is the number of football commercials that Equity decides to purchase and c is the number of comedy commercials purchased, how many of each type of commercial should Equity purchase in order to reach their desired number of female and male viewers?

Begin by writing the system of linear equations, in terms of f and c, needed to solve this problem.

$$
\begin{aligned}
7c + 2f &= 40 \\
2c + 12f &= 80
\end{aligned}
$$

Recall from linear algebra that this system in matrix form, $Ax = b$, looks like

$$
\begin{pmatrix} 7 & 2 \\ 2 & 12 \end{pmatrix} \begin{pmatrix} c \\ f \end{pmatrix} = \begin{pmatrix} 40 \\ 80 \end{pmatrix}
$$

and that there are several techniques for solving this system, including Gaussian elimination and using the inverse of the matrix A, $x = A^{-1}b$. Thus, the solution is

$$
\begin{pmatrix} c \\ f \end{pmatrix} = \begin{pmatrix} \frac{3}{20} & -\frac{1}{40} \\ -\frac{1}{40} & \frac{7}{80} \end{pmatrix} \begin{pmatrix} 40 \\ 80 \end{pmatrix} = \begin{pmatrix} 4 \\ 6 \end{pmatrix}.
$$

Equity should purchase 4 comedy commercials and 6 football commercials in order to reach their desired audience.

∎

Occasionally, it is more important in an application to show that a solution exists than to actually find the solution. In this case, it is possible that there is exactly one solution or infinitely many solutions to the system of linear equations.

The next example will lead us through a problem where the existence of a solution is in question. Again, there are many ways to approach this problem beyond Gaussian elimination, including interpreting the problem in terms of the rowspace and nullspace of a matrix.

Example 3.1.2 *The original Lights Out game is an electronic game that consists of a grid of lights. At the beginning of the game, a random assortment of the lights are turned on to form the initial condition. In a traditional Lights Out game, there are only two states that can be displayed on each of the buttons in the grid, on and off. The goal is to turn each button off by pressing a sequence of buttons. If a button is pressed, it changes its own state and all buttons that are vertically and horizontally adjacent. Figure 3.1 shows the 4×4 grid with the first button and eleventh button pressed, where buttons are numbered left to right starting in the top left corner and ending in the bottom right corner. In this problem, we will assume that all of the lights start on and the goal is to use linear algebra concepts, such as the rowspace and nullspace of a matrix, to determine if a solution exists without necessarily finding a solution.*

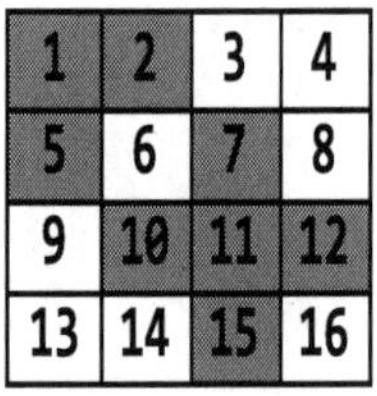

FIGURE 3.1: 4×4 Lights Out with gray buttons off and white buttons on.

We begin by stepping through a few key concepts that are important to setting up this problem.

1. Assuming that all of the lights start on and the goal is to turn them all off, what are the initial and final state vectors?

   In this example, an on button can be represented by a 1 value and an off button can be represented by a 0. Therefore, the initial vector $\vec{i} = \vec{1}$ and the final vector $\vec{f} = \vec{0}$.

2. A 16×1 press vector can also be created to represent the button(s) that are pressed.

We will create a press vector, p, where the jth entry of $\vec{p}$ is a 1 if button j is pressed and a 0 otherwise. Notice that in Figure 3.1, buttons 1 and 11 are pressed, and thus the pressed vector

$$\vec{p} = \begin{pmatrix} 1 \\ 0 \\ 0 \\ 0 \\ 0 \\ 0 \\ 0 \\ 0 \\ 0 \\ 0 \\ 1 \\ 0 \\ 0 \\ 0 \\ 0 \\ 0 \end{pmatrix}.$$

3. How can the change in a button's state, from on to off or off to on, be modeled?

Notice again that in Figure 3.1, when button 1 is pressed buttons 1, 2, and 3 change from on to off. This is because buttons 2 and 3 are adjacent to 1 and we also say button 1 is adjacent to itself in this case. With this in mind, we can create an adjacency matrix. An *adjacency matrix*, A, is a square matrix where entry $A_{i,j} = 1$ when button, vertex, j is connected, or adjacent, to button, vertex, i and 0 otherwise. It should be noted that under these circumstances the matrix A is symmetric. Note that there are applications with connections such that vertex i is connected to vertex j and vertex j is not connected to vertex i. These are similar to one way streets and when these types of connections are introduced symmetry will be lost in your adjacency matrix.

The adjacency matrix for the 4×4 Lights Out grid is

$$A = \begin{pmatrix}
1 & 1 & 0 & 0 & 1 & 0 & 0 & 0 & 0 & 0 & 0 & 0 & 0 & 0 & 0 & 0 \\
1 & 1 & 1 & 0 & 0 & 1 & 0 & 0 & 0 & 0 & 0 & 0 & 0 & 0 & 0 & 0 \\
0 & 1 & 1 & 1 & 0 & 0 & 1 & 0 & 0 & 0 & 0 & 0 & 0 & 0 & 0 & 0 \\
0 & 0 & 1 & 1 & 0 & 0 & 0 & 1 & 0 & 0 & 0 & 0 & 0 & 0 & 0 & 0 \\
1 & 0 & 0 & 0 & 1 & 1 & 0 & 0 & 1 & 0 & 0 & 0 & 0 & 0 & 0 & 0 \\
0 & 1 & 0 & 0 & 1 & 1 & 1 & 0 & 0 & 1 & 0 & 0 & 0 & 0 & 0 & 0 \\
0 & 0 & 1 & 0 & 0 & 1 & 1 & 1 & 0 & 0 & 1 & 0 & 0 & 0 & 0 & 0 \\
0 & 0 & 0 & 1 & 0 & 0 & 1 & 1 & 0 & 0 & 0 & 1 & 0 & 0 & 0 & 0 \\
0 & 0 & 0 & 0 & 1 & 0 & 0 & 0 & 1 & 1 & 0 & 1 & 1 & 0 & 0 & 0 \\
0 & 0 & 0 & 0 & 0 & 1 & 0 & 0 & 1 & 1 & 1 & 0 & 0 & 1 & 0 & 0 \\
0 & 0 & 0 & 0 & 0 & 0 & 1 & 0 & 0 & 1 & 1 & 1 & 0 & 0 & 1 & 0 \\
0 & 0 & 0 & 0 & 0 & 0 & 0 & 1 & 0 & 0 & 1 & 1 & 0 & 0 & 0 & 1 \\
0 & 0 & 0 & 0 & 0 & 0 & 0 & 0 & 1 & 0 & 0 & 0 & 1 & 1 & 0 & 0 \\
0 & 0 & 0 & 0 & 0 & 0 & 0 & 0 & 0 & 1 & 0 & 0 & 1 & 1 & 1 & 0 \\
0 & 0 & 0 & 0 & 0 & 0 & 0 & 0 & 0 & 0 & 1 & 0 & 0 & 1 & 1 & 1 \\
0 & 0 & 0 & 0 & 0 & 0 & 0 & 0 & 0 & 0 & 0 & 1 & 0 & 0 & 1 & 1
\end{pmatrix}.$$

The ones in the product of the adjacency matrix with the press vector, $A\vec{p}$, represent the buttons that change states when the presses in the press vector occur. Now be careful; since we only are working with zeros and ones in this problem, all of our operations will be modulus 2.

4. How can we combine the knowledge of parts 1 through 3 to determine if there is a unique solution to the Lights Out game?

One can think of this problem as the system of equations $A\vec{p} + \vec{i} = \vec{f}$, where we wish to solve for $\vec{p}$. Therefore, the system $A\vec{p} + \vec{i} = \vec{f}$ becomes $A\vec{p} + \vec{1} = \vec{0}$, which is the same as $A\vec{p} = \vec{1}$ since the additive inverse of 1 is 1 modulus 2.

The first thing that you should ask yourself is whether the matrix A is invertible modulus 2. You may wish to use mathematical software to determine if A is invertible modulus 2. If A is invertible modulus 2, then there is a unique solution to this problem. In fact $|A| = 0$ and A is not invertible modulus 2 and thus there is not a unique solution to this Lights Out game. At this point we are uncertain as to whether a solution exists or not.

5. Is there a solution to the 4×4 Lights Out game?

One can certainly answer this question using basic techniques like Gaussian elimination but we will pose a different path. In order to determine if a solution exists for $A\vec{p} \equiv \vec{1} \mod 2$, we are really asking if $\vec{1}$ is in the range of the transformation $A\vec{p}$. Determining if $\vec{1}$ is in the range of $A\vec{p}$ is equivalent to determining if $\vec{1}$ is in the columnspace of A. Since the matrix A is a symmetric matrix, this is the same as determining if $\vec{1}$ is in the rowspace of A.

Recall that the nullspace of A is orthogonal to the rowspace of A. Therefore, one way to determine if a solution exists for $A\vec{p} \equiv \vec{1} \mod 2$ is to determine if

$$\vec{1} \perp \text{nullspace } A.$$

Using mathematical software, we determine that the dimension of the nullspace(A), the nullity, is 2, that is, there are two basis vectors for the nullspace of A,

$$\vec{v_1} = (1,0,1,0,1,1,0,1,0,1,0,0,0,0,1,0,1,0,1,1,0,1,\ 0,1)$$

and

$$\vec{v_2} = (0,1,1,1,0,1,0,1,0,1,1,1,0,1,1,1,0,1,0,1,0,1,1,\ 1,0).$$

Take the Euclidean inner product, dot product, of each of these vectors with $\vec{1}$ to determine if $\vec{1}$ is orthogonal to the nullspace(A).

$$\vec{1}.\vec{v_1} = 12 \equiv 0 \ mod2$$

and

$$\vec{1}.\vec{v_2} = 16 \equiv 0 \ mod2$$

and thus $\vec{1}$ is orthogonal to nullspace(A). Therefore, we can conclude that there is a solution for $A\vec{p} \equiv \vec{1} \mod 2$ and translating back this means there is a solution, and in fact infinitely many solutions, to the 4×4 Lights Out game.

■

Exercises

1. Answer the following questions.

   (a) Rework Example 3.1.1 under the conditions that the company wants to reach 80 million high-income women and 80 million high-income men. In other words, how many comedy commercials and football commercials should Equity Autos purchase in order to reach 80 million high-income women and 80 million high-income men?

   (b) Determine how many of each type of commercial should be purchased to reach c million high-income women and c million high-income men. What must be true of c for your answer to be reasonable physically?

2. Juicemathics wishes to produce three types of mixed juice using cranberry, orange, and pineapple juice. The rows in matrix M represent the three types of juice to be made and the columns represent cranberry, orange, and pineapple juice, respectively. Thus the $(i,j)^{\text{th}}$ entry in M represents how many ounces of each type of juice j to use to produce type i.

$$M = \begin{pmatrix} 4 & 8 & 4 \\ 2 & 10 & 4 \\ 12 & 4 & 0 \end{pmatrix}.$$

(a) Solve the system $Mx = \begin{pmatrix} 16 \\ 16 \\ 16 \end{pmatrix}$ for x.

(b) If each of the juices is produced in 16 ounce bottles, how would you interpret the solution to part 2a?

3. JellyBeanJeans (JBJ) wants to create a line of denim outfits for small children. They make jeans, overalls, and embroidered hats. For a particular size, JBJ needs the materials as presented in Table 3.1.

TABLE 3.1

Material	Jeans	Overalls	Hats
Denim (sq ft)	2	3	1
Metal Brads (number of packs)	1	3	0
Spools of Thread	1	2	6

(a) JBJ has done market research and determined that for every 1 pair of jeans they sell, they sell 2 pairs of overalls and 3 embroidered hats. JBJ would like to find out how many square feet of denim, how many packs of metal brads, and how many spools of thread they need to completely make their items in these relative numbers. If JBJ wanted to produce 1 pair of jeans, 2 pairs of overalls, and 3 embroidered hats, how many of each component should they purchase?

(b) JBJ wants to produce jeans, overalls, and hats exactly in a 1:2:3 proportion but they are limited in how much of each material component they can buy. If JBJ can buy at most 70 spools of thread, how many hats can they make and still maintain both the 1:2:3 proportion and use all of their purchased materials? How many spools should they purchase? (Assume that the amount of the other two components is unlimited.)

(c) If JBJ had exactly 425 square feet of denim, 255 packs of brads, and 289 spools of thread, could they make exactly a whole number of each item? How many of each item could they make?

4. Exponentitown and Logicity are concerned about their local economies. All of the people who work in these two towns also live in these two towns. Those who live and work in the same city spend half of their income in that city each month, saving half of their income. Those who live and work in different cities spend half of their income in each city. 70% of Exponentitown's citizens work in Exponentitown while the other 30% work in Logicity. 60% of Logicity's citizens work in Logicity and the other 40% work in Exponentitown. If c_1 is the population of Exponentitown, c_2 is the population of Logicity, and each individual has the same monthly income of $1000, the income for each city is

$$\begin{array}{ll}\text{(Exponentitown)} & 1000(.50c_1 + .20c_2) \\ \text{(Logicity)} & 1000(.15c_1 + .50c_2)\end{array}$$

(a) We wish to determine the population of Exponentitown and Logicity that will produce economies of 9 and 6 million dollars, respectively. Create a matrix A and vector b so that this problem can be represented as $Ax = b$, where $x = \begin{pmatrix} c_1 \\ c_2 \end{pmatrix}$.

(b) Solve the system $Ax = \begin{pmatrix} 9{,}000{,}000 \\ 6{,}000{,}000 \end{pmatrix}$ using the matrix A from part 4a.

(c) If in fact the two cities were incorrect and instead their income is represented by the matrix $A = \begin{pmatrix} 700 & 300 \\ 350 & 150 \end{pmatrix}$, use the null vector of A^T to argue what Logicity's income must be if Exponentitown's income is $21,000.

5. In this problem, the task is to balance the following chemical equation using a system of linear equations

$$x_1 C_3H_8 + x_2 O_2 \rightarrow x_3 H_2O + x_4 CO_2.$$

(a) Balance the chemical equation above by solving the linear system

$$x_1 \begin{pmatrix} 3 \\ 8 \\ 0 \end{pmatrix} + x_2 \begin{pmatrix} 0 \\ 0 \\ 2 \end{pmatrix} - x_3 \begin{pmatrix} 0 \\ 2 \\ 1 \end{pmatrix} - x_4 \begin{pmatrix} 1 \\ 0 \\ 2 \end{pmatrix} = \begin{pmatrix} 0 \\ 0 \\ 0 \end{pmatrix}.$$

This is equivalent to solving the linear system

$$
\begin{pmatrix}
3 & 0 & 0 & -1 \\
8 & 0 & -2 & 0 \\
0 & 2 & -1 & -2
\end{pmatrix}
\begin{pmatrix}
x_1 \\
x_2 \\
x_3 \\
x_4
\end{pmatrix}
=
\begin{pmatrix}
0 \\
0 \\
0
\end{pmatrix}.
$$

There are infinitely many solutions; find one.

(b) Using your solution to part 5a, if you wanted to have exactly 1.6 moles of water and 1.2 moles of carbon dioxide after the reaction occurs, how many moles of C_3H_8 and O_2 should be present before the reaction takes place?

6. (Input-Output Model) In a particular closed economy, there are three industries producing products p_1, p_2, p_3, respectively, each of which relies on the other. In order to produce one unit of product p_1, $\frac{1}{4}$ of a unit of p_1, $\frac{1}{3}$ of a unit of p_2, and $\frac{1}{4}$ of a unit of p_3 are needed. Similarly, in order to produce one unit of p_2, $\frac{1}{2}$ of a unit of p_1, $\frac{1}{3}$ of a unit of p_2, and $\frac{1}{4}$ of a unit of p_3 are needed, and in order to produce one unit of p_3, $\frac{1}{4}$ of a unit of p_1, $\frac{1}{3}$ of a unit of p_2, and $\frac{1}{2}$ of a unit of p_3 are needed.

From this description, notice that you are actually solving the system

$$
\begin{pmatrix}
\frac{1}{4} & \frac{1}{3} & \frac{1}{4} \\
\frac{1}{2} & \frac{1}{3} & \frac{1}{4} \\
\frac{1}{4} & \frac{1}{3} & \frac{1}{2}
\end{pmatrix}
\begin{pmatrix}
p_1 \\
p_2 \\
p_3
\end{pmatrix}
=
\begin{pmatrix}
p_1 \\
p_2 \\
p_3
\end{pmatrix}.
$$

(a) Determine the percentage of the closed economy that must be made up of p_1 in order to sustain this closed economy.

(b) If 100 units of p_2 must consistently be present in the closed economy, determine how many units of the other two products must remain present in the economy in order to sustain the economy.

7. According to the United Nations International Merchandise Trade Statistics, each of the following countries, China, India, and Singapore, provides large amounts of exports to the others. Table 3.2 shows the units of trade between countries that we will use for this problem.

Assuming a closed economy between these three countries, what ratio of commodities should each country produce in order to keep the economy stable?

TABLE 3.2

Consumption Country	China	India	Singapore
China	0.46	0.2	0.58
India	0.36	0.7	0.13
Singapore	0.18	0.1	0.29

Project[TE]: 5 × 5 Lights Out

1. Under the same rules as presented in Example 3.1.2, determine the adjacency matrix associated with the Lights Out 5 × 5 grid game.

2. If all of the lights start on in the 5 × 5 grid game and the buttons 2, 3, 5, 7-9, 13-17, and 19-22 are pressed, what happens to the lights in the grid?

3. In part 2, you saw one solution to the 5 × 5 Lights Out game. Determine if this is the only solution or if infinitely many solutions exist.

Project[TE]: Forest Growth

61% of the state of North Carolina is forestland. Loblolly pine is the most important commercial timber in the southeastern United States. Over 50% of the standing pine in the southeast is loblolly. This is an easily seeded, fast-growing member of the yellow pine group. On an average site, the loblolly would reach 55-65 feet in 25 years. Thinning of loblolly pine farms should start around 15-20 years.

The goal of this problem is to determine the number of trees to harvest. Let's say that we have planted loblolly pines in our plantation for the past 15 years and thus there are trees at a variety of heights, which we will put into categories, $p_1, p_2, ...p_n$. After 15 years we wish to thin our plantation and thus will harvest trees from each category. The matrix that represents the growth rates is called the growth matrix and is of the form

$$G = \begin{pmatrix} 1-g_1 & 0 & 0 & \cdots & 0 & 0 \\ g_1 & 1-g_2 & 0 & \cdots & 0 & 0 \\ 0 & g_2 & 1-g_3 & 0 & 0 & 0 \\ \vdots & \vdots & \ddots & \ddots & \vdots & \vdots \\ 0 & 0 & \cdots & 0 & 1-g_{n-1} & 0 \\ 0 & 0 & \cdots & 0 & g_n & 1 \end{pmatrix}.$$

1. At the 15 year marker, the beginning of harvesting,

$$x = \begin{pmatrix} x_1 \\ x_2 \\ x_3 \\ x_4 \\ x_5 \end{pmatrix}$$ represents the number of trees in each category. Assuming the

growth is calculated such that the growth matrix G transitioning from one year to the next is

$$G = \begin{pmatrix} 1 & 0 & 0 & 0 & 0 \\ 0.75 & 0.4 & 0 & 0 & 0 \\ 0 & 0.6 & 0.5 & 0 & 0 \\ 0 & 0 & 0.5 & 0.6 & 0 \\ 0 & 0 & 0 & 0.4 & 1 \end{pmatrix},$$

what does Gx represent? From the matrix G you might note that 75% of trees in category 1 move to category 2 in a year (time period). What might the farmer be doing to make the (1,1) entry of G equal to 1?

2. Suppose h_i is the fraction of the ith category that will be harvested at the end of each year, and we let H be the diagonal matrix whose entries are the h_i's. What does HGx represent? What does $Gx - HGx$ represent?

3. Assume that $x_1 = 100$, if $H = \begin{pmatrix} 0 & 0 & 0 & 0 & 0 \\ 0 & 0.1 & 0 & 0 & 0 \\ 0 & 0 & 0.1 & 0 & 0 \\ 0 & 0 & 0 & 0.2 & 0 \\ 0 & 0 & 0 & 0 & 0.8 \end{pmatrix}$; use G and H

to determine how to maintain a sustainable tree farm. Discuss what the 0 in the (1,1) entry of H represents.

3.2   Discrete Stochastic Processes

A *stochastic process* is a process that evolves over time. In fact, a stochastic process is a collection of random variables X_t indexed by time t. In this section, we will focus on *discrete stochastic processes*, which are stochastic process that have discrete time steps, be it days, years, hours, etc. Each random variable X_t, where $t \in \{0,1,\ldots\}$, takes on what we call a *state*. A *Markov chain* is a specific type of discrete stochastic process. In a Markov chain, we can introduce a *transition matrix* where the ij^{th} entry represents the probability of transitioning, in one time step, from state j to state i. We will explain further here with a quick example.

Example 3.2.1 *Let's say you are playing a simple board game where the only rules are to start at the beginning, spot 1, and increase to the next spot on the board based on the number that you get on a fair spinner. The spinner has the numbers 0, 1, and 2, on it, all of which are equally likely outcomes, indicating the number of spaces to move forward. There are 5 spots on the board, 1 through 5, and the winner is the one who lands exactly on spot 5 first, and does not go over. How many turns does each player have to take in order to be confident that there will be a winner?*

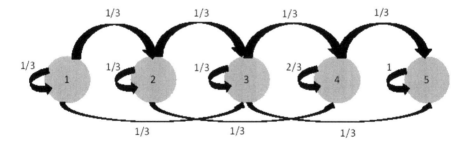

FIGURE 3.2: Example state space diagram.

Figure 3.2 shows a state space diagram of the game described. Notice that if, at time 0, $X_0 =$ spot 1 then X_1, the state at time 1, could be spot 1 with probability $\frac{1}{3}$, spot 2 with probability $\frac{1}{3}$, or spot 3 with probability $\frac{1}{3}$. Also notice that from spot 4, if a player spins a 0 or a 2 then they will stay at spot 4 since a value of exactly 1 is necessary to get to spot 5. Spot 5 is called an *absorbing state* since once a player reaches this spot they do not leave. The probability of going from an absorbing state to that same absorbing state is 1.

From Figure 3.2, we can create a transition matrix, M, where the probability of going from state j to state i is the ij^{th} entry of M.

$$M = \begin{pmatrix} \frac{1}{3} & 0 & 0 & 0 & 0 \\ \frac{1}{3} & \frac{1}{3} & 0 & 0 & 0 \\ \frac{1}{3} & \frac{1}{3} & \frac{1}{3} & 0 & 0 \\ 0 & \frac{1}{3} & \frac{1}{3} & \frac{2}{3} & 0 \\ 0 & 0 & \frac{1}{3} & \frac{1}{3} & 1 \end{pmatrix}.$$

Powers of this transition matrix M can tell us a substantial amount about the long term behavior of the system. For example,

$$M^2 = \begin{pmatrix} \frac{1}{9} & 0 & 0 & 0 & 0 \\ \frac{2}{9} & \frac{1}{9} & 0 & 0 & 0 \\ \frac{1}{3} & \frac{2}{9} & \frac{1}{9} & 0 & 0 \\ \frac{2}{9} & \frac{4}{9} & \frac{1}{3} & \frac{4}{9} & 0 \\ \frac{1}{9} & \frac{2}{9} & \frac{5}{9} & \frac{5}{9} & 1 \end{pmatrix}$$

and thus the probability of going from spot 1 to spot 5 in 2 steps is $\frac{1}{9}$.

If you had to bet on the number of turns it would take you to go from spot 1 to spot 5, how many turns do you think that would be?

The probability of going from spot 1 to spot 5 in 4 steps, $M^4_{5,1}$, is approximately 48% and in 5 steps, $M^5_{5,1}$, is approximately 63%, so it would be wise to bet somewhere between 4 and 5 steps depending on how conservative you tend to be.

■

We have all played games similar to the one in Example 3.2.1, so if we don't stop the game when the winner gets to spot 5 but instead keep playing, will all of the players eventually get to this last spot? We know that in fact they will since spot 5 is an absorbing state and all of the players eventually end up there. But there are Markov chains that do not have an absorbing state. We will see such an example next.

Example 3.2.2 *In Coimbatore forest, Tamil Nadu, India, elephants migrate between three areas, Anaikatti (A), Periya Thadagam (PT), and Nanjundapuram(N). The probability that an elephant will migrate from one region to another on any given day can be found in Table 3.3. What percentage of the Coimbartore elephants will be in each area in the long run?*

TABLE 3.3

		From		
		A	PT	N
	A	.69	.10	.08
To	PT	.09	.75	.26
	N	.22	.15	.66

If M is the transition matrix and 100% of the elephants start in area A on day 0, we can see that

$$M.\begin{pmatrix} 100 \\ 0 \\ 0 \end{pmatrix} = \begin{pmatrix} 69. \\ 9. \\ 22. \end{pmatrix}.$$

That is, after just one day, 9% of the elephants have moved to region PT and 22% have moved to region N while 69% remain in region A.

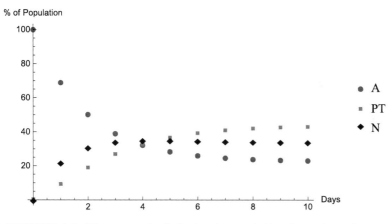

FIGURE 3.3: The percent of elephant population in each region over 10 days.

In Figure 3.3, we can see the percent of the elephant population in each region from day 0 through day 10. Notice that it appears that the three curves are each limiting to a value. For example, it appears that the percent of the elephants in region A in the "long run" is approximately 20%. We call the "long run" state the *steady state*.

The steady state of a Markov chain is affiliated with the eigenvector associated with the eigenvalue of magnitude 1. Recall that to find this eigenvector, we are looking to solve

$$\begin{pmatrix} .69 & .10 & .08 \\ .09 & .75 & .26 \\ .22 & .15 & .66 \end{pmatrix} \begin{pmatrix} x \\ y \\ z \end{pmatrix} = \begin{pmatrix} x \\ y \\ z \end{pmatrix}$$

where x, y, and z are the populations in regions A, PT, and N, respectively.

Using mathematical software, we find that the eigenvector associated with the eigenvalue of magnitude 1 is

$$\vec{v} = \{-0.381784, -0.72871, -0.568527\}.$$

Clearly the values in this eigenvector are not representing percentage populations of elephants. The unit vector, a vector of length 1, in the same direction as $\vec{v}$ will give the steady state percentage populations for each region $\{0.227385, 0.434009, 0.338606\}$.

■

Some discrete stochastic processes are not related to Markov chains; however, looking at powers or eigenvectors of matrices is still important in these problems. Example 3.2.3 is an example of one such process that is used in the PageRank system of Google search.

Example 3.2.3 *For this example, we will assume that the internet has only 4 pages which are linked based on Figure 3.4.*

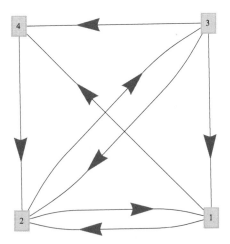

FIGURE 3.4: A 4 page internet system, where each vertex represents a page and each arrow represents a hyperlink.

Google search actually determines an importance ranking based on all of the links related to the pages in the search. [11]

For example, in Figure 3.4, one can think of page 1 as having 2 links to it, from pages 2 and 3. An even stronger way to think of weighting the links is that page 4 puts all of its weight toward a link to page 2 while page 3 splits it weight among three links toward page 1, page 2, and page 4. If x_1, x_2, x_3, and x_4 are the weight that each of the pages has to distribute between its links, respectively, we can think of this as the linear systems

$$x_1 = \frac{x_2}{2} + \frac{x_4}{2}$$
$$x_2 = \frac{x_1}{2} + \frac{x_3}{2}$$
$$x_3 = \frac{x_1}{3} + \frac{x_2}{3} + \frac{x_4}{3}$$
$$x_4 = x_2$$

Under this weighting system we wish to solve the system $M.\vec{x} = \vec{x}$, where

$$\vec{x} = \begin{pmatrix} x_1 \\ x_2 \\ x_3 \\ x_4 \end{pmatrix} \text{ and } M = \begin{pmatrix} 0 & \frac{1}{2} & \frac{1}{3} & 0 \\ \frac{1}{2} & 0 & \frac{1}{3} & 1 \\ 0 & \frac{1}{2} & 0 & 0 \\ \frac{1}{2} & 0 & \frac{1}{3} & 0 \end{pmatrix}.$$ Finding $\vec{x}$ is equivalent to finding

the unit eigenvector associated with the eigenvalue equal to 1. This vector is $\vec{x} = \{0.25, 0.375, 0.1875, 0.1875\}$, which can be interpreted as ranking page 2 as most important. ∎

Exercises

1. ^{TE} If you have a Markov chain with 8 states and transition matrix

$$T = \begin{pmatrix} 0.5 & 0 & 0 & 0 & 0 & 0 & 0 & 0 \\ 0.5 & 0.2 & 0 & 0 & 0 & 0 & 0 & 0 \\ 0 & 0 & 0.1 & 0 & 0 & 0 & 0 & 0 \\ 0 & 0.4 & 0.1 & 0.3 & 0 & 0 & 0 & 0 \\ 0 & 0.5 & 0.2 & 0.3 & 0.1 & 0 & 0 & 0 \\ 0 & 0 & 0.2 & 0.3 & 0.5 & 0.2 & 0 & 0 \\ 0 & 0 & 0.4 & 0.1 & 0.2 & 0.2 & 0.1 & 0 \\ 0 & 0 & 0 & 0 & 0.2 & 0.6 & 0.9 & 1 \end{pmatrix}$$

discuss how you might determine the long term behavior of the matrix simply by the characteristics of the matrix.

2. ^{TE} A certain two year college has been keeping data on their students. Students can be in one of four categories, Drop-Out (D), Year 1, Year 2, or

Graduate (G). The transition matrix for the student body is

$$T = \begin{array}{c} \\ D \\ Y1 \\ Y2 \\ G \end{array} \begin{pmatrix} D & Y1 & Y2 & G \\ 1 & 0.2 & 0.1 & 0 \\ 0 & 0.3 & 0 & 0 \\ 0 & 0.5 & 0.2 & 0 \\ 0 & 0 & 0.7 & 1 \end{pmatrix}.$$

(a) Discuss what the 1 in the (4,4) entry might represent.

(b) If there are currently 100 year 1 students, 200 year 2 students, and 0 students in the other two categories, how many students will graduate in the next 4 years?

(c) In the current model, there is no introduction of new students. If the university wishes to admit the same number of year 1 students as were in the previous year 1 class, while still allowing 50% to move to year 2, determine what the (2,2) entry in T would have to be in order make this adjustment.

3. ^{TE} Math Ninjas is a local community program that offers Saturday martial arts classes for children ages 8 to 12 over the course of 6 months. They recruit from the community and estimate that 10% of the "not-enrolled," eligible new students enroll at the beginning of each month. Children begin in "white belt" status, and as each month passes, the top 30% of the white belt class moves to the "yellow belt" status, but 25% of the class drops out (and is considered to return to a general, "not-enrolled" population). After the end of 1 month in yellow belt status, 40% of the students remain in the program (in the same status) but the rest drop out of the program. Students who drop out of the classes and decide to re-enroll can only begin again as white belt students.

(a) Set up a transition matrix for enrollment in the Math Ninjas program.

(b) The community has 300 children who are 8 to 12 years old. How many white belt students are in the class at the beginning of the program? Approximately how many white belt and yellow belt students are in the class at the beginning of the second month of the program?

4. ^{TE} Consider a game of ladder climbing. There are 5 levels in the game; level 1 is the lowest (bottom) and level 5 is the highest (top). A player starts at the bottom. Each time, a fair coin is tossed. If it turns up heads, the player moves up one rung. If tails, the player moves down to the very bottom. Once

at the top level, the player moves to the very bottom if a tail turns up, and stays at the top if a head turns up. The transition matrix for this game is

$$T = \begin{pmatrix} 0.5 & 0.5 & 0.5 & 0.5 & 0.5 \\ 0.5 & 0 & 0 & 0 & 0 \\ 0 & 0.5 & 0 & 0 & 0 \\ 0 & 0 & 0.5 & 0 & 0 \\ 0 & 0 & 0 & 0.5 & 0.5 \end{pmatrix}.$$

(a) If 10,000 players start at the bottom of the ladder, approximately how many will be on level 3 after 5 plays?

(b) If 10,000 players start at the bottom of the ladder, determine the steady state using eigenvalues and eigenvectors and interpret this in terms of how many players will end up at each level.

5. ^{TE} The pitch array

$$\{64,62,60,62,64,64,64,62,62,62,64,67,67,$$

$$64,62,60,62,64,64,64,64,62,62,64,62,60\}$$

is the start of a well known nursery rhyme. If we assume that similar patterns exists throughout the nursery rhyme, we can see that the probability of going from pitch 64 to pitch 62 is $\frac{5}{11}$, from pitch 64 to pitch 64 is $\frac{5}{11}$, and from pitch 64 to pitch 67 is $\frac{1}{11}$. Complete the following transition matrix and determine the long term behavior, the steady state, of the song.

	60	62	64	67
60			0	
62			$\frac{5}{11}$	
64			$\frac{5}{11}$	
67			$\frac{1}{11}$	

6. ^{TE} In [19], the author describes an interesting stochastic model for college dating. You are told that at the start of orientation 80% of freshman are single, 10% are in a relationship, and 10% are in that ambiguous state termed "it's complicated." After one month there is a 30% chance a single student will remain single, a 20% chance they will enter a relationship, and a 50% chance they will enter the "it's complicated" state. Similarly, if the student starts in a relationship, there is a 20% chance that they are single in a month and a 35% chance that they are in a complicated relationship. Lastly, if the student deems their relationship complicated during orientation, there is a 50% chance that they are single in a month and just a 10% chance that their relationship is not complicated. Given this information, determine the

percent of students that are single after 2 months. What about at the end of freshman year? Do these results depend on the initial condition that 80% were initially single?

7. ^{TE} (Leslie Matrix Problem) The population size of a certain bird species is to be studied. These birds are born in the spring and live at most 3 years. There will be 3 age categories for these birds: Age 0 (born the previous spring), Age 1, and Age 2. Suppose that 20% of Age 0 birds survive to the next spring, and 50% of Age 1 birds survive to become Age 2. Suppose females that are 1 year old produce a clutch of a certain size, of which 3 females are expected to survive to the next breeding season. Females that are 2 years old lay more eggs and 6 females are expected to survive from each of these clutches until the next spring. Use this information to determine how many females are in each age group after 5 years if 100 female birds are initially in each age group. Determine how many female birds would be in each age group after 100 years.

8. ^{TE} (Sensitivity analysis) Using the information from problem 7, how sensitive are the long run results (populations after 100 years) to the initial condition of 100 female birds? To determine sensitivity, change the initial condition slightly and observe what happens to the end behavior with this slight change.

Project^{TE}: A Random Wager

You start on an adventure with 12 dollars in your pocket. Assume that you get one dollar every day thereafter and you put it in your pocket. Each day after collecting your one dollar you take out a random number of dollars from your pocket (all possibilities are equally likely; that is, the probability of pulling one dollar = probability of pulling two dollars = probability of pulling the maximum number of dollars in your pocket) and spend that money. The goal of this project is to estimate how many dollars you will have in your pocket after 3 days.

1. Throughout the 3 day gambling period, there are 13 possibilities, states, for the amount of money that you can have in your pocket. Create a transition matrix where the $(i,j)^{\text{th}}$ entry represents the probability of going from j dollars to i dollars in your pocket.

2. Use the transition matrix that was created in part 1 and the fact that you start with 12 dollars to determine the probability that you will still have only 12 dollars in your pocket after 1 day.

3. Use the transition matrix that was created in part 1 and the fact that you start with 12 dollars to determine the probability that you will still have only 12 dollars in your pocket after 3 days.

4. In order to estimate how much you have in your pocket after 3 days, find the expected value of the 3 day game by finding

$$\sum_{n=0}^{12} n \cdot P(n \text{ dollars after 3 days}).$$

Project^{TE}: Snakes and Ladders

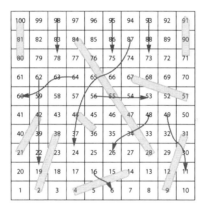

FIGURE 3.5: A 10 × 10 Snakes and Ladders board game.

Using the Snakes and Ladders board shown in Figure 3.5, the goal of this project is to determine, on average, how long it will take to finish the game.

1. If each player rolls a fair 6 sided die on each turn and follows the Snakes and Ladders board shown in Figure 3.5, create a transition matrix where the $(i,j)^{\text{th}}$ entry represents the probability of going from space j to space i in a single turn.

2. If everyone starts on spot 1 at the beginning of the game, what should the initial vector for this game be?

3. Determine the percent of players who have finished the game in 10 turns. 20 turns. 50 turns.

4. Using your findings from part 3, determine on average how long it will take to finish the game.

3.3 Optimization

In Example 3.1.1, Equity Autos determined the number of football and comedy commercials to purchase in order to meet a specific audience, but what if the goal was to reach that audience while minimizing their cost?

In many optimization problems, the goal is to minimize or maximize a function, called the *objective function*, while still satisfying some properties, typically inequalities, called *constraints*. There are a few ways to approach these problems. One way is to graph the constraints to see the *feasibility region*. The feasibility region displays all of the possible solutions that satisfy the constraints. If a maximum or minimum value for the objective function exists, that value would occur at one of the corners of the feasibility region.

Example 3.3.1 *Equity Autos would like the commercials to be seen by **at least** 40 million high-income women and **at least** 80 million high-income men. Let f be the number of football commercials that Equity decides to purchase and c be the number of comedy commercials purchased. If a 1-minute comedy ad costs* $50,000, *and a 1-minute football ad costs* $100,000, *how many of each type of commercial should Equity purchase in order to reach their desired number of female and male viewers and minimize their cost?*

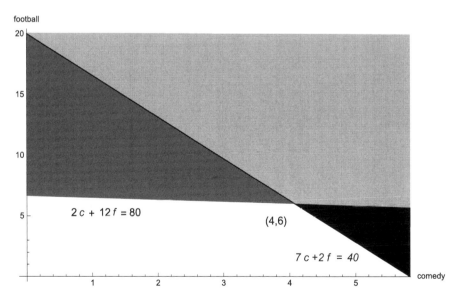

FIGURE 3.6: Constraints for Example 3.3.1 with feasibility region in gray.

Figure 3.6 is a graph of the feasibility region for Example 3.3.1, under the constraints

$$
\begin{aligned}
7c + 2f &\geq 40 \\
2c + 12f &\geq 80 \\
c &\geq 0 \\
f &\geq 0.
\end{aligned}
$$

Notice that the last two constraints are added in order to make sure that the values for the number of comedy ads and football ads make sense in application. In general, we call these constraints *non-negative constraints*.

The objective function $Cost = 50{,}000c + 100{,}000f$ will be minimized at a corner of the feasibility region in Figure 3.6. For example, consider one of the corners of the region, $(4,6)$. This point represents $c = 4$ and $f = 6$, which would result in a $Cost = 800$. Hence if Equity purchases 6 football commercials and 4 comedy commercials then their cost would be \$800,000. If instead Equity purchases 20 football commercials and 0 comedy commercials then their cost would be \$2,000,000 and if they chose to purchase 0 football commercials and 34 comedy commercials their cost would be \$1,700,000. Therefore, if Equity Autos wishes to reach their intended audience while minimizing their cost, they should purchase 6 football commercials and 4 comedy commercials.

∎

Another approach to this type of optimization problem is to use an algorithm called the *simplex method*. The simplex method is an algorithm for solving problems more generally called *linear programming problems*. The simplex method may not be the most efficient method to use in this problem of only 2 variables; however, it is very efficient with larger problems. We will use the information from Example 3.3.1 only for ease of explanation.

In the simplex method, we think about each of our constraints (excluding the non-negative constraints such as $c \geq 0$ and $f \geq 0$) in terms of equalities. In order to do so, we have to add in additional values called *slack variables*. For example, an inequality such as $7c + 2f \geq 40$ can be written as the equality $7c + 2f + s_1 = 40$ where $s_1 \leq 0$ is called a slack variable.

We then organize the information given from the constraints with slack variables included into a matrix (or table) called the *tableau*. The bottom row of the tableau is the coefficients of the objective function written as a homogeneous function with positive cost. For example, in Example 3.3.1, the cost function $Cost = 50{,}000c + 100{,}000f$ would be rewritten as $-50{,}000c - 100{,}000f + Cost = 0$. We will step through the rest of the steps of the simplex method while reworking the Equity Autos problem in Example 3.3.2.

Example 3.3.2 *We begin applying the simplex method to Example 3.3.1 by* *rewriting the constraints as equations with slack variables $s_1 \leq 0$ and $s_2 \leq 0$.*

$$7c + 2f + s_1 = 40$$
$$2c + 12f + s_2 = 80.$$

The tableau can then be constructed with the coefficients of these equations with the coefficients of the objective function in the bottom row.

$$\left(\begin{array}{ccccc|c} c & f & s_1 & s_2 & Cost & \text{constants} \\ \hline 7 & 2 & 1 & 0 & 0 & 40 \\ 2 & 12 & 0 & 1 & 0 & 80 \\ \hline -50000 & -100000 & 0 & 0 & 1 & 0 \end{array} \right).$$

In the simplex method, a *feasible solution* to a problem is one in which all variables are non-negative. One example of a feasible solution is where $c = 3, f = 10$ and $s_1 = -1, s_2 = -46$, which is when Equity Autos purchases 3 comedy commercials and 10 football commercials. In this case, Equity Autos is meeting their viewer needs but not doing so at a minimum cost. We now wish to use the tableau to determine feasible solutions. We begin by picking what is called a *pivot element* in the simplex method.

In order to find the pivot element, determine the column with the most negative number in the last row; this is the pivot column. Then divide each number in the constant column by the corresponding number in the pivot column. The pivot element is the intersection of the column with the most negative number in the last row and the row with the smallest quotient.

In this example, the largest negative number in the last row is in the second column and thus the pivot column will be the second column. The two quotients in the constant column are $\frac{40}{2}$ and $\frac{80}{12}$, the smallest of which is $\frac{80}{12}$ and thus the pivot element is the (2,2) element, 12.

Now use elementary row operations to make this pivot element a 1 and to make the pivot column all zeros above and below the pivot element.

$$\left(\begin{array}{ccccc|c} c & f & s_1 & s_2 & Cost & \text{constants} \\ \hline 7 & 2 & 1 & 0 & 0 & 40 \\ \frac{2}{12} & \boxed{1} & 0 & \frac{1}{12} & 0 & \frac{20}{3} \\ \hline -50000 & -100000 & 0 & 0 & 1 & 0 \end{array} \right).$$

$$\left(\begin{array}{ccccc|c} c & f & s_1 & s_2 & Cost & \text{constants} \\ \hline \frac{20}{3} & 0 & 1 & -\frac{2}{12} & 0 & \frac{80}{3} \\ \frac{2}{12} & \boxed{1} & 0 & \frac{1}{12} & 0 & \frac{20}{3} \\ \hline -50000 & -100000 & 0 & 0 & 1 & 0 \end{array} \right).$$

$$\left(\begin{array}{ccccc|c} c & f & s_1 & s_2 & Cost & \text{constants} \\ \hline \frac{20}{3} & 0 & 1 & -\frac{2}{12} & 0 & \frac{80}{3} \\ \frac{2}{12} & \boxed{1} & 0 & \frac{1}{12} & 0 & \frac{20}{3} \\ \hline -\frac{100000}{3} & 0 & 0 & \frac{25000}{3} & 1 & \frac{2000000}{3} \end{array} \right).$$

In order to find the feasible solution that minimizes the objective function, we continue in this fashion until there are no negative numbers in the last row.

Next determine another pivot element and proceed to get this pivot column in reduced row echelon form. Column 1 will be the pivot column and again to find the pivot element, divide each row constant by the corresponding value in that row and the pivot column, getting values of 4 for row 1 and 40 for row 2. Choosing the smaller value, the (1,1) entry will be the pivot element.

$$\left(\begin{array}{cccccc} c & f & s_1 & s_2 & Cost & \text{constants} \\ \hline \boxed{20/3} & 0 & 1 & -\frac{2}{12} & 0 & \frac{80}{3} \\ \frac{2}{12} & 1 & 0 & \frac{1}{12} & 0 & \frac{20}{3} \\ \hline -\frac{100000}{3} & 0 & 0 & \frac{25000}{3} & 1 & \frac{2000000}{3} \end{array}\right).$$

$$\left(\begin{array}{cccccc} c & f & s_1 & s_2 & Cost & \text{constants} \\ \hline \boxed{1} & 0 & \frac{3}{20} & -\frac{1}{40} & 0 & 4 \\ \frac{2}{12} & 1 & 0 & \frac{1}{12} & 0 & \frac{20}{3} \\ \hline -\frac{100000}{3} & 0 & 0 & \frac{25000}{3} & 1 & \frac{2000000}{3} \end{array}\right).$$

$$\left(\begin{array}{cccccc} c & f & s_1 & s_2 & Cost & \text{constants} \\ \hline \boxed{1} & 0 & \frac{3}{20} & -\frac{1}{40} & 0 & 4 \\ 0 & 1 & -\frac{1}{40} & \frac{7}{80} & 0 & 6 \\ \hline -\frac{100000}{3} & 0 & 0 & \frac{25000}{3} & 1 & \frac{2000000}{3} \end{array}\right).$$

$$\left(\begin{array}{cccccc} c & f & s_1 & s_2 & Cost & \text{constants} \\ \hline 1 & 0 & \frac{3}{20} & \frac{1}{40} & 0 & 4 \\ 0 & 1 & \frac{1}{40} & \frac{7}{80} & 0 & 6 \\ \hline 0 & 0 & 5000 & 7500 & 1 & 800000 \end{array}\right).$$

At this point we see all numbers in the last row are non-negative and thus we know we have come to a point where we have an optimal solution in the constants column. Here $s_1 = s_2 = 0, f = 6$, and $c = 4$ with a minimum cost of roughly \$800,000.

∎

In the next example, we will determine the solution using the feasibility and ask the reader to use the simplex method to determine the solution in Exercise 4.

Example 3.3.3 *Your university has to offer both calculus and statistics classes next semester. There are 35 seats in each calculus class and 30 seats in each statistics class. If there are only 15 classes that can be taught and 480 students that need to take one of these two classes and at least 1 section of each must be taught, determine the number of sections of each class that should be taught in order to minimize the number of empty seats, while also minimizing the number of classes offered (minimizing cost).*

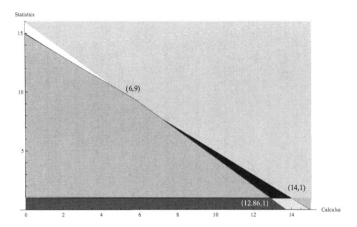

FIGURE 3.7: Feasibility region for Example 3.3.3 in black.

If c represents the number of calculus classes being offered and s represents the number of statistics classes being offered, then the given constraints are

$$
\begin{aligned}
c + s &\leq 15, \\
35c + 30s &\geq 480, \\
c &\geq 1, \\
s &\geq 1,
\end{aligned}
$$

and we wish to minimize objective function $c + s$. Based on the feasibility region in Figure 3.7, the three vertices of the feasibility region are $(6,9), (14,1)$ and approximately $(12.86,1)$. If 6 calculus sections and 9 statistics sections are offered, all 480 students will have a course to take and there will not be any empty seats. If 14 calculus sections and 1 statistics section are offered, then there will be 40 empty seats in these classes. The other option of $(12.86,1)$, rounded to 13 calculus sections and 1 statistics section, could be offered but there would be 5 empty seats.

■

Exercises

1. Given the constraints

$$
\begin{aligned}
x_1 + 2x_2 &\leq 4 \\
4x_1 + 3x_2 &\leq 10 \\
x_1 - x_2 &\geq 1 \\
x_1 &\geq 0 \\
x_2 &\geq 0
\end{aligned}
$$

   plot the feasibility region and determine the values for x_1 and x_2 that maximize the objective function $2x_1 + x_2$.

2. Given the constraints

$$
\begin{aligned}
3x_1 + x_2 + x_3 &\leq 60 \\
x_1 - x_2 + 2x_3 &\leq 10 \\
x_1 + x_2 - x_3 &\leq 20
\end{aligned}
$$

   and the objective function $2x_1 - x_2 + x_3$, set up the initial tableau for the simplex method.

3. Cyborg Robotics has two different products, a robotic claw and a glo spec, that they wish to make. Their goal is to maximize their profit. In order to do so, they have to produce at least one of each product. In addition, due to time restrictions, no more than 20 products can be made and at least as many robotic claws need to be made as glo specs. Determine the constraints and feasibility region.

4. Use the simplex method to find the solution to Example 3.3.3.

5. On a stormy day, you go to the local store and the only foods available are cereal and milk. You know that the storm will last for days and your decision about how much of each food to buy is made entirely on dietary and economic considerations.

   Using the information from Table 3.4, determine how many units of cereal and milk you should buy if you wish to meet all of your dietary needs and only have $20.

TABLE 3.4

	Per unit-cereal	Per unit-milk	Min. requirements
Units of carbohydrates	3	1	8
Units of vitamins	4	3	19
Units of protein	1	3	7
Unit cost	$2	$4	

6. ^{TE} A toy brick company is selling two different types of character packs, knights of the round table and halloween ghouls and goblins. Each pack comes with 4 characters. The knights of the round table require 4 head/body/leg packages, 12 accessories, and 3 weapons while the halloween ghouls and goblins pack includes 4 head/body/leg packages, 18 accessories, and 1 weapon. The halloween pack is more popular among children and thus is sold for $6 per pack while the knights pack only costs $4 per pack. The brick company gets in a large amount of 4 head/body/leg packages each week but only 100 accessories and 20 weapons. Determine how many of each pack type should be produced each week in order maximize revenue.

7. ^{TE} The toy brick company from Exercise 6 has decided to add one more character to their ghouls and goblin pack. With this addition, the ghouls and goblin pack requires 5 head/body/leg packages, 18 accessories, and 2 weapons. The brick company receives 25 head/body/leg packages. Using this new information as well as the information from Exercise 6, determine how many of each pack type should be produced each week in order to maximize revenue.

Project: Trenitalia Train Optimization

Trenitalia is looking to purchase two new types of trains, the Red Arrow and the Silver Arrow, to help them with their transportation needs between Nice, France, and Florence, Italy. The seating capacity of the Red Arrow is 600 customers and of the Silver Arrow is 250. The demand for each route can be seen in Table 3.5.

1. If r is the number of Red Arrow trains purchased and s is the number of

TABLE 3.5

Round trip route	Average fare	Demand (customers)
Nice-Florence	Red Arrow $120/person Silver Arrow $100/person	2000
Florence-Rome	Red Arrow $80/person Silver Arrow $60/person	3000

FIGURE 3.8: Map of train routes.

Silver Arrow trains purchased to make the round trip journey from Nice-Florence, and Trenitalia is willing to purchase up to 5 trains to meet these demands, determine the constraints of this problem.

2. Determine the number of Red Arrow trains and Silver Arrow trains that Trentitalia should purchase if they wish to minimize the number of empty seats on this route.

3. It costs $10 per km to operate a Red Arrow train and $8 per km to operate a Silver Arrow train. Determine the number of Red Arrow trains and Silver Arrow trains that Trenitalia should purchase for their Nice-Florence round trip route in order to maximize profit if the distance on this route is 900 km.

4. Trenitalia also wishes to purchase trains for its round trip Florence-Rome journey, which is 500 km. The fare and demand for this journey on each type of train can be found in Table 3.5. If Trenitalia is willing to purchase

up to 8 trains to meet the demands of the 2 routes (Florence-Rome and Nice-Florence), how many of each type should they purchase for each route in order to maximize their profits?

Project: Roller Coaster

The local amusement park is designing a new roller coaster, called the Maximus. In order to run safely, the weight and height of the front, middle, and back cars of the ride must follow the regulations found in Table 3.6. Additional regulations on the Maximus can be found in Table 3.7.

TABLE 3.6
Individual Car Restrictions

Car	Average Weight(lbs)	Average Height(in)	Max number of riders
Front	140	63	14
Middle	160	68	6
Back	240	76	10

TABLE 3.7
Combined Ride Restrictions

Minimum Riders	Maximum Weight(lbs)	Minimum Height(inches)
7	2200	550

1. If only 1 rider is in the front car and there must be at least 1 rider in each of the other cars, how many riders should be in each of the other cars in order to maximize the number of riders?

2. If there are 2 riders in the middle car and there must be at least 1 rider in each of the other cars, how many riders should be in each of the other cars in order to maximize the number of riders?

3. If only 1 rider is in the back car and there must be at least 1 rider in each of the other cars, how many riders should be in each of the other cars in order to maximize the number of riders?

4. The park drops the Maximus' maximum weight regulation, but instead sets a regulation that the ratio of number of riders to average weight must be larger in the back car than in the front car. If there are 2 riders in the middle car and there must be at least 1 rider in each of the other cars, how many riders should be in each of the other cars in order to maximize the number of riders?

3.4   Linear Techniques for Data Analysis - Least Squares Regression

Many of us think of least squares regression as a statistical modeling tool and some of you may have even seen a derivation of least squares in multivariate calculus. In fact, the most well known derivation of least squares regression can be seen through linear algebra techniques.

Recall that when we wish to fit a least squares regression line $\hat{y} = A\hat{x}$ to a set of data, $\{\{x_1,y_1\},\{x_2,y_2\},\ldots,\{x_n,y_n\}\}$,

$$A = \begin{pmatrix} x_1 & 1 \\ x_2 & 1 \\ \vdots & \vdots \\ x_n & 1 \end{pmatrix}, \hat{x} = \begin{pmatrix} a \\ b \end{pmatrix}, y = \begin{pmatrix} y_1 \\ y_2 \\ \vdots \\ y_n \end{pmatrix},$$

and $\hat{y}$ is an estimation of y. More explicitly, $\hat{y} = A\hat{x}$ creates a set of new

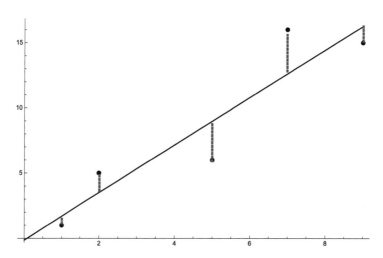

FIGURE 3.9: Figure of data with regression line and residuals shown.

points $\{\{x_1,\hat{y_1}\},\{x_2,\hat{y_2}\},\ldots,\{x_n,\hat{y_n}\}\}$ that are on the regression line. Notice in Figure 3.9, the error between the actual data point, y, and the point on the line affiliated with the same x value, $A\hat{x}$, called the *residual*. In a least squares regression model, we wish to to minimize the sum of the squared residuals. In linear algebra, this is equivalent to minimizing $(y - A\hat{x})^T(y - A\hat{x})$.

In calculus, you have learned that minima (and optima in general) usually occur at the critical values. In this minimization problem, we will take the derivative with respect to $\hat{x}$ and thus

$$\frac{d}{d\hat{x}}(y - A\hat{x})^T(y - A\hat{x}) = \frac{d}{d\hat{x}}y^Ty - y^TA\hat{x} - \hat{x}^TA^Ty + \hat{x}^TA^TA\hat{x}$$

$$= \frac{d}{d\hat{x}}y^Ty - 2y^TA\hat{x} + (A\hat{x})^TA\hat{x}$$

$$= -2A^T(y - A\hat{x}).$$

Recall that in order to find the critical values we set $-2A^T(y - A\hat{x}) = 0$, which is the same as $A^Ty - A^TA\hat{x} = 0$, and solve for $\hat{x}$. Therefore the $\hat{x}$ that minimizes the error is

$$\hat{x} = (A^TA)^{-1}A^Ty.$$

Example 3.4.1 *Table 3.8 shows the state population as of July 2015 [20] and the number of Zika viruses associated with travel in 2015-2016 (as of July 13, 2016)[12] for 10 states in the United States.*

TABLE 3.8
Population versus Number of Zika
Cases

Arizona	6,828,065	7
Kansas	2,911,641	5
Louisiana	4,670,724	7
Michigan	9,922,576	13
Mississippi	2,992,333	8
North Carolina	10,042,802	18
Ohio	11,613,423	25
South Carolina	4,896,146	13
Utah	2,995,919	5
Vermont	626,042	3

From the sample of 10 states represented in Table 3.8 and Figure 3.10, we can see that there appears to be a positive linear behavior. That is, as the population size of states increase, the number of travel related cases of Zika also increases. Assuming that this linear relationship exists, we can find a least squares regression line to model this data.

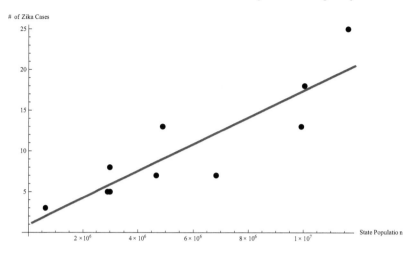

FIGURE 3.10: Population of states versus number of travel related cases of Zika with least squares regression line.

We begin by defining

$$A = \begin{pmatrix} 6{,}828{,}065 & 1 \\ 2{,}911{,}641 & 1 \\ 4{,}670{,}724 & 1 \\ 9{,}922{,}576 & 1 \\ 2{,}992{,}333 & 1 \\ 10{,}042{,}802 & 1 \\ 11{,}613{,}423 & 1 \\ 4{,}896{,}146 & 1 \\ 2{,}995{,}919 & 1 \\ 626{,}042 & 1 \end{pmatrix} \text{ and } y = \begin{pmatrix} 7 \\ 5 \\ 7 \\ 13 \\ 8 \\ 18 \\ 25 \\ 13 \\ 5 \\ 3 \end{pmatrix}.$$

$(A^T A)^{-1} A^T y = \begin{pmatrix} 1.64 \times 10^{-6} \\ .96 \end{pmatrix}$. Therefore, the least squares regression line that you see in Figure 3.10 is $\hat{y} = (1.64 \times 10^{-6})x + .96$.

∎

Notice that in Example 3.4.1, the slope of the regression line is very small and thus you might question whether another curve would be a better fit of your data. We can use similar techniques to those shown in Example 3.4.1 to fit higher degree polynomials to a set of data.

Example 3.4.2 *You might decide that in fact a second degree polynomial, $\hat{y} = ax^2 + bx + c$, would better fit the data found in Table 3.8. In this case, you would* *have three parameters with* $\hat{x} = \begin{pmatrix} a \\ b \\ c \end{pmatrix}$ *and*

$$A = \begin{pmatrix} (6{,}828{,}065)^2 & 6{,}828{,}065 & 1 \\ (2{,}911{,}641)^2 & 2{,}911{,}641 & 1 \\ (4{,}670{,}724)^2 & 4{,}670{,}724 & 1 \\ (9{,}922{,}576)^2 & 9{,}922{,}576 & 1 \\ (2{,}992{,}333)^2 & 2{,}992{,}333 & 1 \\ (10{,}042{,}802)^2 & 10{,}042{,}802 & 1 \\ (11{,}613{,}423)^2 & 11{,}613{,}423 & 1 \\ (4{,}896{,}146)^2 & 4{,}896{,}146 & 1 \\ (2{,}995{,}919)^2 & 2{,}995{,}919 & 1 \\ (626{,}042)^2 & 626{,}042 & 1 \end{pmatrix}.$$

In this case, the least squares second degree polynomial fit is

$$\hat{y} = (1.32 \times 10^{-13})x^2 - (4.72 \times 10^{-8})x + 4.66$$

and can be see in Figure 3.11.

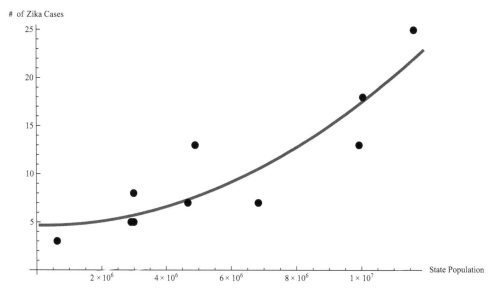

FIGURE 3.11: Population of states versus number of travel related cases of Zika with least squares quadratic function.

Again, here you should notice that the coefficient related to the x^2 term is very small and thus you may decide a linear fit is more reasonable. So how do we decide which of these two models is the best model for our data? It is important at this point to think about the error associated with each of your models; however, before discussing error analysis we should also think about whether your models are reasonable based on the application.

When you are fitting a curve to your data, as we have done here, you may want to think about some of the following regarding assumptions.

- If you have chosen a curve that takes on negative y values, does it make sense that this dependent variable could be negative or should the curve really just take on positive values for the dependent variable?

- If you have chosen a curve whose limit approaches infinity as x, the independent variable, approaches infinity, is it possible for the dependent variable to take on infinitely large values or is it more practical for the dependent variable to reach some capacity? If a capacity is more appropriate, then a different model, a *logistic model*, which cannot be found using the techniques in this chapter, is more appropriate in this case.

- If the model you have chosen has a term with a very small coefficient, is that term essential to the validity of the model?

- If residuals related to your model have similar characteristics, such as all positive (or negative) in an area, or may not appear to be random in nature, then perhaps a more complicated model is appropriate.

Once you have identified reasonable potential curves to fit your data, you can consider which ones produce the minimum overall error. Error is measured in many different ways. There is a *maximum error*

$$||y - \hat{y}||_\infty = maximum_i |y_i - \hat{y}_i|$$

which is the highest residual over the entire dataset. In Example 3.4.1, the largest residual between the least squares regression line and the dataset occurs when $x = 6828065, y = 7$, and $\hat{y} = 12.17$, and thus $||y - \hat{y}||_\infty = 5.17$. For the second degree polynomial, found in Example 3.4.2, the largest error occurs when $x = 4896146, y = 13, \hat{y} = 7.60936$ and $||y - \hat{y}||_\infty = 5.39$. Based on a lower maximum error value of 5.17, we may argue that the linear regression is a more appropriate model for the Zika dataset than the quadratic model. However, the maximum error does not tell us much about the overall fit of the data.

Another way to measure error in a dataset of size n is with the l_2 error,

$$||y - \hat{y}||_2 = \sqrt{\left(\sum_{i=1}^{n} |y_i - \hat{y}_i|^2 \right)}.$$

If you were to calculate the l_2 error for the model given in Example 3.4.1, you would find that the l_2 error when using the least squares regression line is still a little larger, 9.76583, than that with the quadratic fit, 8.82312, found in

Example 3.4.2. The downfall of the l_2 error is that it does not take into account the size of the values of the dependent variable. A small error for a dataset of small dependent variable values could be viewed very differently than an error of the same size for a dataset of large values.

If you wish to integrate the size of the data into the error, use the relative l_2 error

$$\frac{||y - \hat{y}||_2}{||y||_2} = \frac{\sqrt{(\sum_{i=1}^{n} |y_i - \hat{y}_i|^2)}}{\sqrt{(\sum_{i=1}^{n} |y_i|^2)}}.$$

For the linear and quadratic models in Examples 3.4.1 and 3.4.2, the relative l_2 errors is approximately 25% and 23% respectively. For this data set, all three error measurements are fairly similar for both the linear and quadratic models, leaving room for argument as to which is best.

Exercises

1. Players of a local soccer team believe that the more times they practice kicking with their non-dominant foot the more accurate they will be at shooting on goal. The team statistician decides to create a model to predict a player's accuracy percentage based on the number of times the player practiced kicking with their non-dominant foot. Explain why a least squares regression line is not the best model to use with this data.

2. The least squares regression line

$$\widehat{\text{support}} = 260(\text{facebook interest}) - 169$$

relates the percent of Trump supporters in 2016 polls with facebook interest in Trump (by state).

   (a) In the key state of Michigan, 87% of facebook users were showing an interest in Trump online. Using this information and the least squares regression line, what should the polls have been predicting about Trump's support in the 2016 election?

   (b) Prior to the election, 67% of facebook users from D.C. were showing interest in Trump online. Using the regression line, determine what the polls should have been predicting about the support for Trump in the 2016 election.

   (c) Discuss the issues with using a linear regression model to represent the behavior of this data.

3. [TE] Education data for 20 states can be seen in Table 3.9.

(a) Use the linear methods described in this chapter to fit a least squares regression line using funds given to predict the number of college graduates.

(b) One of the states has the goal of having 750,000 college graduates. Use your regression from part 3a to determine how much should be given to the local educational agencies in that state. (Round your answer to the nearest million dollars.)

TABLE 3.9
Funding for Local Education [21] vs College Graduates [22]

State	Funds for Educational Agencies	College Graduates
Alabama	221,717,283	189,259
Connecticut	116,021,685	190,044
Florida	775,553,867	816,946
Idaho	57,316,420	66,871
Illinois	663,983,900	798,362
Kansas	104,106,288	157,023
Maine	50,093,161	53,357
Michigan	498,675,460	434,937
Minnesota	148,615,160	350,909
Montana	45,468,962	48,068
New York	1,104,439,248	1,302,196
North Carolina	417,088,622	462,802
Oklahoma	156,295,251	152,441
Oregon	140,324,584	194,831
Pennsylvania	544,122,676	654,558
South Carolina	225,765,906	205,399
Texas	1,320,732,434	1,143,206
Virginia	243,579,617	477,103
Wisconsin	208,476,889	291,007
Wyoming	33,060,090	24,821

4. ^{TE} Table 3.10 shows life expectancy at birth for males. Life expectancy at birth is the predicted average life expectancy for a baby born in a given year.

(a) Use this data to create a least squares regression line using year to predict the male life expectancy at birth.

(b) Tim was born in 1973. According to the regression from part 4a, what is Tim's life expectancy in years and months?

TABLE 3.10

United States Male Life Expectancy at Birth per Year [23]

year	1960	1964	1968	1972	1976	1980	1984	1988	1992
life exp.	66.6	66.8	66	67.4	69.1	70	71.1	71.4	72.3
year	1996	2000	2004	2008	2012				
life exp.	73.1	74.1	75	75.6	76.4				

5. ^{TE} Determine how to use the least squares regression methods discussed in this chapter to fit a third degree polynomial to the data in Table 3.8, do so, and discuss how this model compares to the linear and quadratic models in the example.

6. ^{TE} In order to use the least squares method in this section, the problem needs to be able to be formulated as a linear equation. Consider the data presented in Table 3.8 with the potential model $\hat{y} = be^{ax}$.

   (a) We would like to create a regression similar to $\hat{y} = A\hat{x}$, but an exponential function is not linear. Note that

$$\begin{aligned} \hat{y} &= be^{ax} \\ \ln(\hat{y}) &= \ln(be^{ax}) = \ln(b) + \ln(e^{ax}) \\ &= ax + \ln(b) \end{aligned}$$

and we can then consider a linear problem $\ln(\hat{y}) = A\hat{x}$ where $\hat{x} = \begin{pmatrix} a \\ \ln(b) \end{pmatrix}$. Find A for this converted equation.

   (b) Using the A from part 6a, find the least squares fit for $\hat{y} = be^{ax}$.

Project^{TE}: Latitude versus Zika

Thus far, we have fit regression models to the data in Table 3.8 using state population to predict the number of travel related Zika cases. In Table 3.11, we see the average latitude per state. In this exercise, we wish to use two independent variables, population size and latitude, in order to predict the number of travel related Zika cases in a state.

1. Create a matrix with columns representing population, latitude, and the coefficient (as described in this section) and use this matrix to determine the coefficients $a, b,$ and c for the model

$$\widehat{\text{Zika cases}} = a \text{ population } + b \text{ latitude } + c.$$

TABLE 3.11

Latitudinal Data [16]

Arizona	34.2932
Kansas	38.4847
Louisiana	31.0891
Michigan	44.3471
Mississippi	32.7250
North Carolina	35.5437
Ohio	40.2912
South Carolina	33.9357
Utah	39.3238
Vermont	44.0751

2. In Section 1.3, you learned about transformations of data. When analyzing data, you may sometimes find that a transformation, on data representing the dependent or independent variable, better fits the model. Create a matrix with columns representing $ln(population)$, a logarithmic transformation of the population data, *latitude*, and the coefficient and use this matrix to determine the coefficients a, b, and c for the model

$$\widehat{\text{Zika cases}} = a \, \ln(\text{population}) \, + b \, \text{latitude} \, + c.$$

3. Discuss which of the models created in parts 1 and 2 is a better fit to the data based on error analysis.

Project^{TE}: Male Life Expectancy with an Upper Limit

In Exercises, part 4a, a regression line was determined to predict life expectancy data for U.S. men as presented in Table 3.10. However, life expectancy may have an upper limit [13]. In this project, we will consider fitting the data from Table 3.10 using curves that limit to a particular value. We will not only use least squares techniques but also consider how to incorporate what we would expect about a reasonable curve to describe life expectancy as a function of birth year.

We will begin by considering a curve with the form

$$\hat{y} = \frac{cx}{x + k} \, . \tag{3.1}$$

An example of this type of curve is shown in Figure 3.12. Note that the function levels off to a single output value as $x \to \infty$. Although the function in (3.1) is nonlinear, the equation can be rewritten in a form that can be viewed as a linear problem. In order to do so, recall that we can consider our data as a list

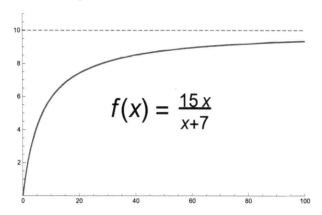

FIGURE 3.12: An example of a function of the form (3.1) with $c = 10$ and $k = 7$.

of points $\{x_i, y_i\}$ and our approximations to the fit will be $\hat{y}_i$.

$$\hat{y}_i = \frac{cx_i}{x_i + k}$$

$$= \frac{c}{1 + \frac{k}{x_i}}$$

$$\hat{y}_i\left(1 + \frac{k}{x_i}\right) = c$$

$$\frac{1}{c}\left(1 + \frac{k}{x_i}\right) = \frac{1}{\hat{y}_i}$$

$$\frac{k}{c}\left(\frac{1}{x_i}\right) + \frac{1}{c} = \frac{1}{\hat{y}_i} \tag{3.2}$$

which can be viewed as a linear problem $A\hat{x} = \hat{z}$ where $\hat{x} = \begin{pmatrix} \frac{k}{c} \\ \frac{1}{c} \end{pmatrix}$ and $\hat{z}_i = \frac{1}{\hat{y}_i}$.

1. What needs to be true about all the data $\{x_i, y_i\}$ so that the above linear problem can be considered? (Hint: Let $x_1 = 10$ correspond to the year 1960.)

2. Find A for the constructed linear problem above using the data in Table 3.10.

3. Find $\hat{x}$ that minimizes the error of the constructed linear problem.

4. Use your answer to part 3 to state a curve of the form in (3.1) that approximates the life expectancy data in Table 3.10.

5. What is the maximum life expectancy for U.S. men according to your answer to part 4?

Considering established knowledge is important when creating a model to represent an application. The study in [13] suggests that the maximum human lifespan may be around 115 years. Additionally, when considering lifespan in the 20th century, a lower limit, m, may also be reasonable. Now we will consider a curve of the form

$$\hat{y} = \frac{cx}{x+k} + m \qquad (3.3)$$

which has the same shape as the curve in Figure 3.12 but shifted up the y-axis m units.

6. What is the limiting value of Equation (3.3)? In other words, what happens to $\hat{y}$ as $x \to \infty$?

7. Based on Table 3.10, we could reasonably choose $m = 60$ as a lower limit for our regression curve. If $m = 60$, what should c be so that the limiting value of the curve in (3.3) is 115?

8. Construct a linear problem to find a regression of the form (3.3) using similar steps to those shown in (3.2) with $m = 60$ and the c value from the previous question. (Hint: you will have a column vector for A and will only be finding k.)

9. Rework part 8 with a different m value that you feel produces a better fit. Plot all three of your regressions (from part 4, part 8, and the current problem) on the same graph along with the data from Table 3.10. Label all curves clearly.

10. The regression found in part 4 had an upper limit to life expectancy and had two values that we could use to fit to data (c and k). The other two regressions in your plot involved using least squares to find fewer values (one, just k) to fit data, but our choices helped find a nice fitting curve with application considerations in mind. Summarize how well (or not well) each of the three curves in your plot predicts the data in Table 3.10, and comment on the overall process of finding an appropriate curve to model the given data.

3.5   Linear Techniques for Data Analysis - Seriation

In some modeling problems, you are not asked to solve something but rather are asked to determine similarities, or differences, between records or even clusters of common records. In this section, we explore records that are binary in nature. That is, we look at data that has or does not have particular attributes. The techniques that are discussed in this section can be applied to a variety of fields including archeology [18], folklore studies [9], semantics, and music genomics [8].

In all of these problem, objects are being compared and the corresponding matrix will be composed of 0s and 1s. Each row will represent an object and each column will represent an attribute or trait. If object i has trait j then the $(i,j)^{\text{th}}$ entry of the matrix will be a 1. If the object is lacking that trait then the entry will be a 0. We will begin with an example of how to set up such a problem into a matrix.

Example 3.5.1 *Imagine that as you grew up, your parents wanted to keep track of popular culture during the course of your life. For every year of your life, it would be a family ritual to bury a time capsule in your background containing those items which dominated popular culture during that year. For example, in 2013 Frozen was one of the most popular kids movies and thus your parents may have included an Elsa doll in the time capsule that was buried in 2013. Furby was also a favorite toy of 2013 so it too may have made your time capsule for 2013. Elsa, from Frozen, was still very popular in 2014, but Furby lost its popularity. In this case, your parents would still bury a Elsa doll in 2014 but not include a Furby. In this example, we discuss how to organize data into a binary matrix based on the presence, or absence, of particular traits in order to prepare to order or cluster the data based on similarities.*

Table 3.12 shows all items in time capsules for 2013, 2014, 2015, and 2016.

We can use this information to create a matrix A whose rows are the time

TABLE 3.12
Time Capsule Items

	Frozen	Furby	Nerf	Lego	X-Box	Anki Drive	Minion	FitBit
2013	1	1	0	1	0	0	0	0
2014	1	0	1	1	1	1	0	0
2015	1	0	1	1	1	0	1	0
2016	0	0	1	1	0	0	1	1

capsules (by year in increasing order) and the columns are the items that may be in the time capsule, in the order from Table 3.12. Since Frozen shows up in 2013 and 2014 we will put a 1 in the (1,1) and (2,1) entry to represent the presence of this artifact. Furby will only be in the 2013 time capsule and thus a 1 can be seen in the (1,2) entry but a 0 is in the (2,2) entry. Continuing in this manner with the data from Table 3.12 we get

$$A = \begin{pmatrix} 1 & 1 & 0 & 1 & 0 & 0 & 0 & 0 \\ 1 & 0 & 1 & 1 & 1 & 1 & 0 & 0 \\ 1 & 0 & 1 & 1 & 1 & 0 & 1 & 0 \\ 0 & 0 & 1 & 1 & 0 & 0 & 1 & 1 \end{pmatrix}.$$

■

The matrix in this example is set up in the ideal world, where we know the chronological order of the capsules. In the real world, you are typically given a set of objects and you are asked to put some order to them.

Now let's assume after 4 years you decide to go out and dig up the time capsules but you have no recollection of which capsule represents which year. You would like to use the presence of, or absence of, an artifact as an indicator of year and thus you must order the rows of matrix A. Notice here that A follows the time capsules through 4 years and each time capsule has the possibility of holding any of 8 artifacts. We do assume that once an artifact goes away, that is, goes from 1 to 0, that it will not come back.

The first technique that we will present in order to try to order records is called *seriation*. In general, seriation is a method of dating or putting records in chronological order. The seriation methods presented here work toward finding an ordering that minimizes some "cost" function.

In all seriation methods, the rows of the original matrix are permuted and a cost is affiliated with each permutation. A permutation can be written in one of two ways: (123) and $\begin{pmatrix} 1 & 2 & 3 \\ 2 & 3 & 1 \end{pmatrix}$ represent the same permutation where 1 goes to 2, 2 goes to 3, and 3 goes to 1. Similarly, the permutation (213), or $\begin{pmatrix} 1 & 2 & 3 \\ 3 & 1 & 2 \end{pmatrix}$, sends 1 to 3, 3 to 2, and 2 to 1. A permutation of the form (1)(23), or $\begin{pmatrix} 1 & 2 & 3 \\ 1 & 3 & 2 \end{pmatrix}$, sends 1 to itself, 2 to 3 and 3 to 2. Each of these permutations is on 3 objects. In total there are $3! + 3 + 1 = 10$ permutations of 3 objects; however, they are not all unique. For example, the permutation (123) and the permutation (231) each sends 1 to 2, 2 to 3, and 3 to 1. The unique permutations on 3 objects are, (123), (213), (1)(2)(3), (12)(3), (1)(23), and (2)(13).

We use these permutations to permute the rows of the original matrix. So a permutation of (123) on a matrix $A = \begin{pmatrix} 1 & 1 & 0 & 1 \\ 1 & 0 & 1 & 1 \\ 1 & 0 & 1 & 1 \end{pmatrix}$ can be done by multiplying matrix A on the left by $P = \begin{pmatrix} 0 & 1 & 0 \\ 0 & 0 & 1 \\ 1 & 0 & 0 \end{pmatrix}$,

$$PA = \begin{pmatrix} 1 & 0 & 1 & 1 \\ 1 & 1 & 0 & 1 \\ 1 & 0 & 1 & 1 \end{pmatrix}.$$

The first cost function that we will look at is related to the number of consecutive ones in the matrix A [10]. Note that if you wish for two consecutive records, or artifacts, as in Example 3.5.1, to be similar (or from the same time period) then they should have similar traits or attributes. Since a one in the data or record represents the presence of a trait and a zero represents the absence of that trait, we wish to maximize the number of consecutive ones or minimize the number of zeros between ones, which we will call *embedded zeros*, in order to put an ordering to our records.

Example 3.5.2 *Let's assume that when the 4 time capsules were dug up, the binary timecapsule-artifact matrix created is the unordered,*

$$A = \begin{pmatrix} 0 & 0 & 1 & 1 & 0 & 0 & 1 & 1 \\ 1 & 0 & 1 & 1 & 1 & 0 & 1 & 0 \\ 1 & 1 & 0 & 1 & 0 & 0 & 0 & 0 \\ 1 & 0 & 1 & 1 & 1 & 1 & 0 & 0 \end{pmatrix} \begin{matrix} \text{Capsule 1} \\ \text{Capsule 2} \\ \text{Capsule 3} \\ \text{Capsule 4} \end{matrix}.$$

We can add up the number of embedded zeros in order to order the time capsules.

Column 3 has 1 embedded zero and in total there are 2 embedded zeros. Permuting rows of the matrix, we change the chronological order of the artifacts and will most likely change the number of embedded zeros as well. Permuting rows 1 and 2

$$\begin{pmatrix} 0 & 1 & 0 & 0 \\ 1 & 0 & 0 & 0 \\ 0 & 0 & 1 & 0 \\ 0 & 0 & 0 & 1 \end{pmatrix} A = \begin{pmatrix} 1 & 0 & 1 & 1 & 1 & 0 & 1 & 0 \\ 0 & 0 & 1 & 1 & 0 & 0 & 1 & 1 \\ 1 & 1 & 0 & 1 & 0 & 0 & 0 & 0 \\ 1 & 0 & 1 & 1 & 1 & 1 & 0 & 0 \end{pmatrix} \begin{matrix} \text{Capsule 2} \\ \text{Capsule 1} \\ \text{Capsule 3} \\ \text{Capsule 4} \end{matrix}$$

still has an embedded zero cost of 3. Permuting rows 1 and 4 produces

$$\begin{pmatrix} 0 & 0 & 0 & 1 \\ 0 & 1 & 0 & 0 \\ 0 & 0 & 1 & 0 \\ 1 & 0 & 0 & 0 \end{pmatrix} A = \begin{pmatrix} 1 & 0 & 1 & 1 & 1 & 1 & 0 & 0 \\ 1 & 0 & 1 & 1 & 1 & 0 & 1 & 0 \\ 1 & 1 & 0 & 1 & 0 & 0 & 0 & 0 \\ 0 & 0 & 1 & 1 & 0 & 0 & 1 & 1 \end{pmatrix} \begin{matrix} \text{Capsule 4} \\ \text{Capsule 2} \\ \text{Capsule 3} \\ \text{Capsule 1} \end{matrix}$$

which has an embedded zero cost of 2.

The permutation (1423) produces the original matrix seen in Example 3.5.1 and thus has 0 embedded zeros. This permutation would give a capsule ordering of Capsule 3, Capsule 4, Capsule 2, Capsule 1. Keep in mind that seriation techniques do not guarantee a unique ordering with minimum cost. Permuting rows 3 and 4 also gives 0 embedded zeros.

∎

It is also important to note that in real world situations, unlike that in Example 3.5.2, it is unlikely to find a permutation that will produce 0 embedded zeros. In this case, the goal should be to minimize the number of embedded zeros.

Another cost function of interest is related to minimizing the number of dissimilarities between consecutive records. This cost is found by creating a similarity matrix

$$S = AA^T.$$

Notice that the $(i,j)^{\text{th}}$ entry of S tells you how many traits records i and j have in common and the entries on the diagonal of S tell you how many traits each individual record possesses.

One can create a dissimilarity matrix $D = N - S$, where each entry of N is equal to the total number of traits. If D is an $n \times n$ matrix, then the cost can also be calculated by

$$\sum_{i=1}^{n-1} D_{i,i+1} + D_{n,1}.$$

Again the goal is to minimize this cost and thus we wish to find a permutation of the rows of A that minimizes the number of dissimilarities.

Example 3.5.3 *Using the matrix A from Example 3.5.2, we can order the time capsules by finding a permutation of the rows of A to minimize the dissimilarity cost function.*

$$S = AA^T = \begin{pmatrix} 4 & 3 & 1 & 2 \\ 3 & 5 & 2 & 4 \\ 1 & 2 & 3 & 2 \\ 2 & 4 & 2 & 5 \end{pmatrix},$$

$$D = \begin{pmatrix} 8 & 8 & 8 & 8 \\ 8 & 8 & 8 & 8 \\ 8 & 8 & 8 & 8 \\ 8 & 8 & 8 & 8 \end{pmatrix} - \begin{pmatrix} 4 & 3 & 1 & 2 \\ 3 & 5 & 2 & 4 \\ 1 & 2 & 3 & 2 \\ 2 & 4 & 2 & 5 \end{pmatrix} = \begin{pmatrix} 4 & 5 & 7 & 6 \\ 5 & 3 & 6 & 4 \\ 7 & 6 & 5 & 6 \\ 6 & 4 & 6 & 3 \end{pmatrix}.$$

From here, we calculate the dissimilarity cost

$$\sum_{i=1}^{3} D_{i,i+1} + D_{4,1} = 5 + 6 + 6 + 6 = 23.$$

Permuting rows 1 and 2 in matrix A from Example 3.5.2 produces a dissimilarity matrix

$$D = \begin{pmatrix} 8 & 8 & 8 & 8 \\ 8 & 8 & 8 & 8 \\ 8 & 8 & 8 & 8 \\ 8 & 8 & 8 & 8 \end{pmatrix} - \begin{pmatrix} 5 & 3 & 2 & 4 \\ 3 & 4 & 1 & 2 \\ 2 & 1 & 3 & 2 \\ 4 & 2 & 2 & 5 \end{pmatrix} = \begin{pmatrix} 3 & 5 & 6 & 4 \\ 5 & 4 & 7 & 6 \\ 6 & 7 & 5 & 6 \\ 4 & 6 & 6 & 3 \end{pmatrix}$$

and thus a dissimilarity cost of $5+7+6+4=22$. The permutation (1423) produces the original matrix seen in Example 3.5.1 with

$$D = \begin{pmatrix} 8 & 8 & 8 & 8 \\ 8 & 8 & 8 & 8 \\ 8 & 8 & 8 & 8 \\ 8 & 8 & 8 & 8 \end{pmatrix} - \begin{pmatrix} 5 & 2 & 2 & 4 \\ 2 & 3 & 1 & 2 \\ 2 & 1 & 4 & 3 \\ 4 & 2 & 3 & 5 \end{pmatrix} = \begin{pmatrix} 3 & 6 & 6 & 4 \\ 6 & 5 & 7 & 6 \\ 6 & 7 & 4 & 5 \\ 4 & 6 & 5 & 3 \end{pmatrix}$$

and thus a dissimilarity cost of 22. ∎

Again there is not a unique permutation that produces a minimum dissimilarity cost. One strategy for finding the best ordering might be to find an ordering that does a good job of minimizing both the embedded zeros cost and the dissimilarity cost.

Exercises

1. The dissimilarity matrix affiliated with record-attribute matrix

$$A = \begin{pmatrix} 0 & 1 & 0 & 0 & 0 & 1 & 1 \\ 1 & 1 & 0 & 0 & 1 & 0 & 1 \\ 1 & 0 & 1 & 1 & 1 & 1 & 1 \end{pmatrix}$$

is

$$D = \begin{pmatrix} 4 & 5 & 5 \\ 5 & 3 & 4 \\ 5 & 4 & 1 \end{pmatrix}.$$

(a) Determine the dissimilarity cost associated with the matrix A using the matrix D above.

(b) If rows 1 and 2 in matrix A are permuted, determine the dissimilarity matrix and the dissimilarity cost associated with this new ordering. (The new ordering would be Record 2, Record 1, Record 3).

2. A matrix is called **Pre-Petrie** if a permutation on the rows can be found such that no embedded zeros exists. Which of the following matrices are Pre-Petrie?

(a) $\begin{pmatrix} 1 & 0 & 1 \\ 0 & 1 & 1 \\ 1 & 1 & 0 \end{pmatrix}$

(b) $\begin{pmatrix} 1 & 1 & 1 & 0 \\ 1 & 0 & 1 & 1 \\ 0 & 1 & 0 & 0 \end{pmatrix}$

(c) $\begin{pmatrix} 1 & 1 & 1 & 0 & 1 \\ 1 & 0 & 1 & 1 & 0 \\ 1 & 1 & 0 & 0 & 1 \\ 1 & 0 & 0 & 1 & 1 \end{pmatrix}$

3. You are given a problem with 14 records to analyze with several traits each. Discuss the computational problems with using seriation techniques to order these records.

4. ^{TE} Your job is to analyze 4 pieces of music for similar attributes (crescendo, decrescendo, staccato, and portamento) and to determine which songs are most closely related. The binary data in Table 3.13 was generated based on numerical data from each song, where a 0 represents a less than average value related to the attribute and a 1 represents a more than average existence.

TABLE 3.13

	Cres.	Decres.	Stac.	Porta.	Time of Song
Like a Virgin	0	0	0	0	0
Billie Jean	1	1	0	0	1
Wanna Dance	1	0	1	1	1
All of You	0	1	1	1	1

The song-attribute matrix is

$$A = \begin{pmatrix} 0 & 0 & 0 & 0 & 0 \\ 1 & 1 & 0 & 0 & 1 \\ 1 & 0 & 1 & 1 & 1 \\ 0 & 1 & 1 & 1 & 1 \end{pmatrix}.$$

(a) Determine the unique permutations of the rows of A.

(b) For each permutation identified in part 4a, determine the number of embedded zeros.

(c) Determine the ordering of the songs based on one of the permutations that gives the minimum number of embedded zeros from part 4b. Those songs that are closest together in the ordering will be "most related."

5. Determine the similarity and dissimilarity matrices affiliated with the song-attribute matrix, A, from Exercise 4.

6. For the permutations $(1)(2)(3)(4)$ and $(1)(234)$, determine the similarity matrix, dissimilarity matrices, and dissimilarity cost. Use this information to determine which songs are most similar.

Project^{TE}: **World Cup Advertising**

The attributes of five countries' Coca Cola advertisements in the 2014 World Cup were studied based on Pollay's list of appeals in advertising [17]. The data for the appeals that were present in each of these advertisements can be seen in Table 3.14.

TABLE 3.14

	Independence	Popularity	Status	Nurturance	Durability	Productivity
Argentina	0	1	0	1	1	0
Germany	0	1	1	0	0	0
Italy	0	1	0	1	0	0
United States	0	1	0	1	1	1

1. Using Table 3.14, create an advertisement-attribute matrix, A.

2. Determine all of the unique permutations of rows of the matrix A in part 1.

3. For each of the permutations in part 2, determine the number of embedded zeros.

4. For each of the permutations in part 2, determine the similarity matrix, dissimilarity matrix, and dissimilarity cost.

5. Use your results from parts 3 and 4 to determine which advertisements are similar in relation to the presence (or absence) of appeals.

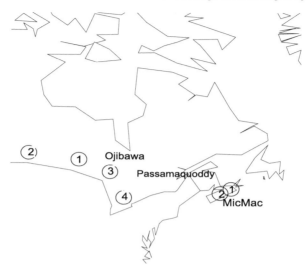

FIGURE 3.13: Geographical Map of Woodland Area

Project^{TE}: The Star Husband Tale

The Native American folklore tale "The Star Husband Tale" has been told throughout history and has generated 86 versions of this tale in North America with 135 traits in total [14]. Table 3.15 represents a tribe-trait binary matrix of 11 traits from 7 of these versions from the Woodlands area. Use the data from Table 3.15 and seriation methods to determine how the Star Husband Tale may have migrated through these tribes. A map of the general location of these tribes can be seen in Figure 3.13 for reference.

TABLE 3.15

Tribe-Trait data for Woodlands Area

Ojibwa 1	1	1	1	1	0	0	0	1	1	0	1
Ojibwa 2	0	1	1	0	0	0	0	1	0	0	1
Ojibwa 3	1	1	1	0	1	1	0	1	0	0	1
Ojibwa 4	1	0	0	0	0	0	0	0	0	1	1
MicMac 1	1	1	0	0	1	0	1	1	0	0	1
MicMac 2	1	1	0	0	1	0	1	1	0	0	0
Passamaquoddy	1	1	1	0	1	0	1	1	0	0	1

1. Use the data from Table 3.15 to create a tribe-trait matrix, A, and determine the number of embedded zeros and dissimilarity cost for matrix A.

2. Permute rows 5 and 7 in matrix A to create matrix B_1 and determine the number of embedded zeros and dissimilarity cost for matrix B_1.

3. Permute rows 6 and 7 in matrix B_1 to create matrix B_2 and determine the number of embedded zeros and dissimilarity cost for matrix B_2.

4. Use your results from parts 1 through 3 and the map in Figure 3.13 to make some final conclusions about how the Star Husband Tale may have migrated through these tribes.

3.6   Linear Techniques for Data Analysis - Singular Value Decomposition

The technique presented in this section is related to a data analysis technique typically used for reducing noise. This technique, called *singular value decomposition (SVD)*, can also be used to compare records visually in two dimensions when significant differences between records exist.

As you should have noticed in Section 3.5, typically matrices related to records versus traits are not square matrices. We can determine if a square matrix is diagonalizable and note that a resulting diagonal matrix contains the eigenvalues of the original matrix on the main diagonal. However, not all matrices are diagonalizable, or even square for that matter. In this case, you may look at singular value decomposition (SVD). If A is an $m \times n$ matrix, its *singular values*, σ_i, are the square roots of the eigenvalues for the matrix $A^T A$.

Like other decompositions, SVD is a way to write the original $m \times n$ matrix, A, as a product of matrices, in this case

$$A = UWV^T,$$

where U is an $m \times m$ orthogonal matrix whose columns are orthonormal eigenvectors of AA^T, V is an $n \times n$ orthogonal matrix whose columns are orthonormal eigenvectors of $A^T A$, and W is an $m \times n$ matrix where $W_{i,i} = \sigma_i$.

Example 3.6.1 *We can find the singular value decomposition of the 3×5 matrix*

$$A = \begin{pmatrix} 0 & 1 & 0 & 0 & 0 \\ 1 & 0 & 0 & 0 & 1 \\ 0 & 0 & 1 & 1 & 0 \end{pmatrix}.$$

$$AA^T = \begin{pmatrix} 1 & 0 & 0 \\ 0 & 2 & 0 \\ 0 & 0 & 2 \end{pmatrix}.$$

The eigenvalues of a diagonal matrix such as AA^T are the entries on the main diagonal and thus the eigenvalues of AA^T are 2 (multiplicity 2) and 1. Recall that to find a basis for the eigensystem corresponding to eigenvalue $\lambda = 2$, we wish to solve for $\vec{x}$ in the equation $AA^T \vec{x} = 2\vec{x}$. There are infinitely many solutions to this linear system of the form $(0,y,z)$ and thus a basis for this eigensystem corresponding to $\lambda = 2$ is $\{(0,1,0),(0,0,1)\}$. Similarly one can find a basis for the eigensystem to $\lambda = 1$ to be $\{(1,0,0)\}$. Construct U with these 3 eigenvectors as its columns, forming an orthogonal matrix, $U = \begin{pmatrix} 0 & 0 & 1 \\ 0 & 1 & 0 \\ 1 & 0 & 0 \end{pmatrix}.$

The eigenvalues of $A^T A$ are 2 (multiplicity 2), 0 (multiplicity 2), and 1 and thus

$$W = \begin{pmatrix} \sqrt{2} & 0 & 0 & 0 & 0 \\ 0 & \sqrt{2} & 0 & 0 & 0 \\ 0 & 0 & 1 & 0 & 0 \end{pmatrix}.$$

A basis for the eigensystem corresponding to $A^T A$'s eigenvector $\lambda = 2$ is $\{(1,0,0,0,1),(0,0,1,1,0)\}$; however, this is not an orthonormal basis. These two vectors do form an orthogonal set; in order to create an orthonormal basis we will divide each vector by $\frac{1}{\sqrt{2}}$. Similarly, we will use the orthonormal basis $\{(-\frac{1}{\sqrt{2}},0,0,0,\frac{1}{\sqrt{2}}),(0,0,-\frac{1}{\sqrt{2}},\frac{1}{\sqrt{2}},0)\}$ for the eigensystem corresponding to $A^T A$'s eigenvector $\lambda = 0$. The eigenvector for $\lambda = 1$ is $\{(0,1,0,0,0)\}$. Using these 5 vectors as the columns of V, such that $A = UWV^T$,

$$V = \begin{pmatrix} 0 & \frac{1}{\sqrt{2}} & 0 & -\frac{1}{\sqrt{2}} & 0 \\ 0 & 0 & 1 & 0 & 0 \\ \frac{1}{\sqrt{2}} & 0 & 0 & 0 & -\frac{1}{\sqrt{2}} \\ \frac{1}{\sqrt{2}} & 0 & 0 & 0 & \frac{1}{\sqrt{2}} \\ 0 & \frac{1}{\sqrt{2}} & 0 & \frac{1}{\sqrt{2}} & 0 \end{pmatrix}.$$

∎

As mentioned at the beginning of the section, there are many applications related to SVD. SVD is commonly used to reduce noise in data. When given noisy data, the dominant (larger) singular values are related to the actual data and the less significant singular values are related to the noise. Figure 3.14 shows an example of where this technique was applied.

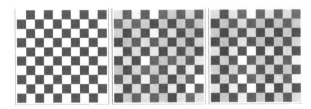

FIGURE 3.14: Grid with no noise, grid with noise of maximum magnitude .01, and grid after SVD using 2 singular values.

Like seriation that was discussed in Section 3.5, singular value decomposition is used to look for similarities between records. Singular value decomposition breaks down data into linear independent components. The data records are represented as the row vectors of the matrix U. If we use singular value decomposition to visualize records of data, with more than 2 values, in the Cartesian plane, we will focus on 2 of those components, and thus 2 columns of U. By reducing the number of singular values to the two largest, we can assign an

(x,y) coordinate to each record.

If we limit the number of singular values to two, the approximate matrix

$$\hat{A} = \hat{U}\hat{W}\hat{V}^T,$$

where $\hat{W}$ is a 2×2 diagonal matrix with main diagonal entries equal to the two largest singular values of $A^T A$. If the original matrix A is an $m \times n$ matrix, then $\hat{U}$ and $\hat{V}$ are $m \times 2$ and $2 \times n$ matrices. The columns of $\hat{U}$ still form an orthonormal set and span(columns of $\hat{U}$)=span(1st two columns of U) and the columns of $\hat{V}$ form an orthonormal set and span(columns of $\hat{V}$)=span(1st two columns of V).

Example 3.6.2 *We can approximate the matrix A from Example 3.6.1, with SVD limited to the two largest singular values.*

$$\hat{A} = \begin{pmatrix} 0 & 0 \\ 0 & 1 \\ 1 & 0 \end{pmatrix} \begin{pmatrix} \sqrt{2} & 0 \\ 0 & \sqrt{2} \end{pmatrix} \begin{pmatrix} 0 & \frac{1}{\sqrt{2}} \\ 0 & 0 \\ \frac{1}{\sqrt{2}} & 0 \\ \frac{1}{\sqrt{2}} & 0 \\ 0 & \frac{1}{\sqrt{2}} \end{pmatrix}.$$

∎

Now let's see how to apply singular value decomposition, with this type of technique, to visualize the data from Example 3.5.1 in the Cartesian plane.

Example 3.6.3 *The matrix A in Example 3.5.1 has singular values $\{3.44096, 1.63674, 1.32939, 0.844726\}$. In the matrix A, each of the 4 records has 8 values, related to the 8 possible traits. Limiting the singular value decomposition to only the two largest singular values, each row of U focuses on two components of each record and can be used to visualize each record in the Cartesian plane.*

$$\hat{U} = \begin{pmatrix} 0.424994 & -0.781465 \\ 0.618297 & -0.0760132 \\ 0.318971 & 0.481529 \\ 0.579092 & 0.389442 \end{pmatrix}$$

and Figure 3.15 shows a graph of the 4 capsules based on these coordinates.

Judging closeness, similarity, based on distance between points on Figure 3.15, one can see that Capsule 1 is closest to Capsule 2. Capsule 4 is closest to Capsule 3 and in general the ordering Capsule 1, Capsule 2, Capsule 3, Capsule 4 would be the ordering that minimizes the distance between capsules using SVD.

∎

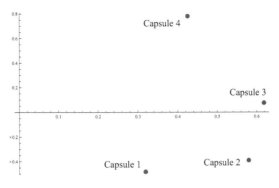

FIGURE 3.15

Exercises

1. Find the singular value decomposition of $\begin{pmatrix} 1 & 1 & 0 \\ 1 & 1 & 0 \\ 0 & 0 & 1 \end{pmatrix}$.

2. Use the two largest singular values of $A = \begin{pmatrix} 1 & 1 & 0 \\ 1 & 1 & 0 \\ 0 & 0 & 1 \end{pmatrix}$ and the singular value decomposition method to approximate A.

3. [TE] Find the singular values of the song-attribute matrix in Exercise 4 in Section 3.5. Round the answers to 2 decimal places.

4. [TE] The attributes of five countries' Coca Cola advertisements in the 2014 World Cup were studied based on Pollay's list of appeals in advertising [17]. The data for the appeals that were present in each of these advertisements can be seen in Table 3.14 in Section 3.5.

    (a) Create an advertisement-appeals matrix and decompose this matrix using SVD.

    (b) Calculate the singular value decomposition of the matrix from part 4a, with the 2 largest singular values.

    (c) Graph the points representing each advertisement found in U from part 4b to determine which countries' advertisements are most similar with regard to the presence (or absence) of appeals.

Project^{TE}: Singular Value Decomposition and Text Analysis

We will be analyzing the following quotes for clusters of similarities using SVD.

Quote 1: Education is the most powerful weapon which you can use to change the world. (Mandela)
Quote 2: Tell me and I forget. Teach me and I remember. Involve me and I learn. (Franklin)
Quote 3: Intelligence is the ability to adapt to change. (Hawkings)
Quote 4: Live as if you were to die tomorrow. Learn as if you were to live forever. (Gandhi)
Quote 5: It always seems impossible until it's done. (Mandela)
Quote 6: If you don't like something, change it. If you can't change it, change your attitude. (Angelou)
Quote 7: The measure of intelligence is the ability to change. (Einstein)
Quote 8: Education is the most powerful weapon which you can use to change the world.(Mandela)

1. Create a matrix, A, with quotes as the rows and the possible words (all words without repeats in the 8 quotes) representing the columns. $A_{i,j}$ = Frequency of word j in quote i. For example, if "the" is the word represented in column 3, then $A_{1,3} = 2$.

2. Calculate the singular value decomposition of A using only the first two singular values.

3. Graph the points representing each quote found in U from part 2 to determine which quotes are most similar with regard to the presence (or absence) of certain words.

Project^{TE}: Genetic Analysis

The Mouse Atlas Project [15] aims to produce a collection of various mouse tissues representing different developmental stages, ranging from embryonic stem cells to post-natal day. In this project, we will use singular value decomposition to order a small collection of this genetic data based on similarities in the sequence of DNA-bases. The 10 digit DNA sequence tags for six distinct tissue samples can be seen in Table 3.16.

1. Create a binary matrix where each row represents a tissue sample and the entries in the matrix represent each DNA-base in the sequence for the respective sample. Notice that each sequence in Table 3.16 has only T, thymine, and A, adenine, and thus a 0 or 1 can be assigned to these DNA-bases, respectively.

TABLE 3.16

Sample 1	T	T	A	T	A	T	T	A	A	A
Sample 2	A	A	A	T	A	A	A	T	T	T
Sample 3	T	A	A	T	A	A	A	A	A	T
Sample 4	T	T	A	A	A	T	T	A	T	A
Sample 5	T	T	T	T	T	T	T	T	T	T
Sample 6	A	A	A	A	A	A	A	A	A	A

2. Create a singular value decomposition of the matrix created in part 1 using the largest singular value (use just 1 singular value) and use the 1×6 matrix U to order the samples.

3. Samples 3 and 6 are from the mouse heart ventricle, samples 2 and 4 are spleen tissue, and samples 1 and 5 are samples from the pancreas. Discuss how these common tissue locations are supported by the ordering from part 2.

4. If the tissue sequences also contained C, cytosine, and G, guanine, discuss how you would go about constructing your matrix prior to the singular value decomposition.

Chapter Synthesis: Independent Investigation

1. (Revisiting Problem 7 from Section 3.1): According to the United Nations International Merchandise Trade Statistics, each of the following countries, China, India, and Singapore, provides large amounts of exports to the others. Table 3.2 shows the units of trade between countries that were used for a previous exercise. Let's say you are from Singapore and your government would like to consume more of its own products. You and your team have investigated the exports of your country and have realized that you can change your within country consumption to 35% (or 0.35 in the table). You believe you can continue to export the same amount to China and India, but you will necessarily import less in total. You need to create a recommendation to your government about how much to import from China and from India. Use different values for import percentages and determine what ratio of commodities each country should produce to keep the economy stable. Develop a report with recommendations to your government based on your investigation.

2. ^{TE} (Revisiting Problem 7 from Section 3.1): According to the United Nations International Merchandise Trade Statistics, each of the following countries, China, India, and Singapore, provides large amounts of exports to the others. Table 3.2 shows the units of trade between countries that were used for a previous exercise. Assume that you are part of a new (but very small) country that has formed in Asia and wants to be part of this export/import structure. Your new country's land happens to be rich in a newly found chemical, Amazingium, which is quickly becoming necessary as a unit of trade to other countries. However, your small country also still needs many items from other countries. You have discovered that China, India, and Singapore will each use 5% of their consumption solely in Amazingium, and your country will only use 15% of its own products in its consumption. Reform Table 3.2 to now include your new country, and adjust the numbers of the table to reflect changes in the consumption overall. Justify your choices and determine what ratio of commodities each country should produce to keep the economy stable. You have considerable power to choose how your country imports products but you should have consumption numbers for other countries that are similar to their current values. Develop a report with recommendations to your country based on your investigation.

3. ^{TE} (Revisiting Example 3.2.2 in Section 3.2) In Coimbatore forest, Tamil Nadu, India, elephants migrate between three areas, Anaikatti (A), Periya Thadagam (PT), and Nanjundapuram (N). The probability that an ele-

phant will migrate from one region to another on any given day can be
found in Table 3.3.

(a) Use techniques from a previous section to develop a list of data points
for the elephant population in each area over 10 days if all elephants
start in area A. (Note this data is presented in a previous figure.) Then
develop a continuous function to model the data for each of the three
regions. Use derivatives to investigate how the populations are changing.

(b) Use techniques from a previous section to develop a list of data points
for the elephant population in each area over 10 days if all elephants
start in area PT. Then develop a continuous function to model the
data for each of the three regions. Use derivatives to investigate how
the populations are changing.

(c) Use techniques from a previous section to develop a list of data points
for the elephant population in each area over 10 days if all elephants
start in area N. Then develop a continuous function to model the data
for each of the three regions. Use derivatives to investigate how the
populations are changing.

(d) Summarize your work and discuss the differences and similarities of
your work in parts 3a-3c.

4. ^{TE} Section 3.5 contains a project entitled "Latitude versus Zika." Rework
this project by creating at least one new model (different from the two presented in the original project) that predicts Zika cases based on latitude
and population. Use error analysis to compare your new model (or models)
to both models from the original project. Include a discussion as to which
model best fits the data.

5. ^{TE} **Linear Modeling Techniques to Predict Stock Price Behavior:**
Table 3.17 shows the first of the month stock prices for 6 stocks of interest.
The goal of the following exercises is to use this data and the topics presented
in this chapter to model some of the behaviors of the prices of these stocks.

(a) Plot both the AAPL and MSFT data and discuss the similarities in
the data.

(b) Use techniques from Chapter 3.4 to fit a 4th degree polynomial to the
AAPL data.

TABLE 3.17

Handle	Jan	Feb	March	April	May	June	July	Aug	Sept	Oct	Nov	Dec	Yearly Average
AMZN	757	833	852	580	664	527	726	757	775	841	785	745	750
AAPL	116	127	139	107	93	97	93	105	107	112	110	109	100
TWTR	16	17	15	16	14	15	17	19	20	18	17	19	17
MSFT	62	64	65	54	49	51	50	56	57	57	59	60	51
NFLX	132	141	142	103	92	101	95	91	98	103	125	117	101

(c) Find a transformation, similar to that learned in Section 1.3, of the function found in part 5b to model the MSFT data.

(d) Use the models found in parts 5b and 5c to discuss the months in which you would expect to see the highest prices for these two stocks.

(e) Create a binary stock-month matrix where a stock receives a 1 in a given month if the price at the beginning of the month is at or above the yearly average for that stock, and a 0 otherwise.

(f) Use seriation techniques from Section 3.5 to determine which stocks are similar in relation to price trends based on the matrix created in part 5e.

(g) Use singular value decomposition to visualize the data in the matrix in part 5e in the Cartesian plane and to determine which stocks are similar in relation to price trends.

(h) We have seen how to use polynomials, seriation, and singular value decomposition to make conclusions about stock price behaviors. Discuss how you would go about using Markov chains to make additional predictions about stock prices and apply these techniques to make some conclusions.

(i) You have $2500 to buy stock shares and you wish to diversify your stock portfolio by purchasing stocks from at least 2 of the stocks in Table 3.17. Discuss what you would buy, when you would buy them, and when you would sell them in order to maximize profit if you were required to sell all of these stocks at the same time. You may assume that the same behaviors are seen each year, so your knowledge of the data in Table 3.17 can be used for future purchases.

4

Modeling with Differential Equations

Goals and Expectations

The following chapter is written toward students who have completed an introductory differential equations course. However, examples within the chapter have been written to include guidance for solving differential equations with computer software. Because of the potential use of technology to solve differential equations, the chapter could also be used by students who have not yet completed a differential equations course, but are somewhat familiar with computational software. Note that the *Mathematica* commands and *MATLAB* code presented are intended solely as examples, and not as instructional components for the respective software.

Goals:

- Section 4.1 (Introduction and Terminology): To introduce important terminology that is used when discussing differential equation models.

- Section 4.2 (First Order Differential Equations): To construct and solve models using first order differential equations.

- Section 4.3 (Systems of First Order Differential Equations): To construct and simulate models using a system of first order differential equations.

- Section 4.4 (Second Order Differential Equations): To construct and solve models utilizing second order differential equations.

4.1   Introduction and Terminology

In calculus, we learned that the derivative $\frac{dx}{dt}$ is the ratio of the differential dx to the differential dt and that this derivative represents the change in the variable x with respect to the change in the variable t. For this derivative, the variable x is the dependent variable, and the variable t is an independent variable. Furthermore, in an introductory differential equations course, we learned that a differential equation is any equation that contains a derivative, or differentials. Because differential equations can be used to define derivatives, or rates of change, they can be used to model virtually any quantity or system that changes over time. In this chapter, we'll discuss how to build and solve a variety of differential equations models.

When creating a differential equations model, the goal is to derive a mathematical representation of how a quantity of interest is changing. This quantity is the dependent variable in the equation, and is referred to as a *state variable* or *state* of the model. The independent variable from the differential equation usually represents a physical attribute such as time or distance. Other parameters or constants that are present within the model are referred to as parameters.

Recall that a differential equation that is coupled with a set of additional constraints on the dependent variable and its derivatives at a particular input (called initial conditions) is called an initial value problem. When constructing a mathematical model using a differential equation, it is desirable to create a model that has a single unique solution rather than a family of solutions. Because of this aspect of the modeling process, the majority of the differential equations models that we will construct or encounter will be initial value problems. While modeling with differential equations, we may also encounter boundary value problems, which consist of a differential equation with additional constraints on the dependent variable and/or its derivatives at different inputs (called boundary conditions).

Within the next sections of this chapter, we will discuss how to construct and solve differential equations models. One aspect of the construction process that we won't discuss very much is the importance of making units match. When formulating a differential equation that will be used to represent physical processes, we must keep in mind that the units of the model must match the units of the physical process. For example, consider the following equation that is part of an SVIR disease model,

$$\frac{dS}{dt} = -\alpha SI + \nu S - \delta S,$$

where S represents the susceptible people in the population and I represents the infected people in the population. Consider the left hand side of the equation, $\frac{dS}{dt}$. This represents the change in the susceptible population over time, and would have units of persons per day. This means that every term on the right hand side of the equation must also be in terms of persons per day. Consider

$-\alpha SI$. Since S and I are both in terms of persons, α must have units of per person per day $\frac{1}{\text{persons}\cdot\text{day}}$. Similarly, ν and δ must both have units of per day $\left(\frac{1}{\text{day}}\right)$.

4.2 First Order Differential Equations

The most basic differential equation model is one that is composed of a single first order differential equation. This kind of model is used to represent the dynamics of a single state variable.

Constructing a First Order Differential Equation Model

The most basic type of first order differential equation model is constructed by identifying all of the various interactions or events that cause a change in the state variable of interest. To create a model for the state variable of interest, we will set the derivative of the state variable with respect to time $\frac{dx}{dt}$ equal to the sum of the mathematical expressions of the influences which have been identified. Note that if an influence would cause an increase in the state variable, then that term would be added, and any influences which detract from the state variable would be subtracted.

It is often useful to draw a schematic of the state variable of interest along with the different interactions that affect the state variable. Typically, in this kind of schematic, the state variable is represented by a box, and the different changes are represented by arrows that point into or out of the box to represent whether the state variable is increased or decreased, respectively, by the given interaction. In the following examples, we will discuss the process of how to create a model using a graphical schematic.

Example 4.2.1 *Stephen is a new college student and is living on his own for the first time. He has decided to establish a checking account at his local bank to help him keep up with his financial responsibilities. He opens the account with a $4000 initial deposit. Every month he deposits a paycheck from his part-time job at the lollipop factory, as well as a check from his grandmother. Additionally, his checking account earns interest that is proportional to the balance of his account. The expenses which he will pay with money from his account include a cell phone bill, rent, and a donation to the local dog rescue. Draw a schematic to represent the changes in Stephen's checking account, and use the schematic to derive a differential equation model for his account balance.*

We'll begin by drawing a schematic that represents the monthly changes in Stephen's bank account. We draw a box to represent our state variable (account balance), and we'll draw arrows pointing in and out of the box to represent the various deposits and withdrawals. Let A represent the amount of money in the account. As stated above, there are three monthly deposits which must be dealt with, so our schematic will need three arrows pointing into the box. Let G represent the monthly deposit from Grandma, P represent Stephen's monthly paycheck, and I represent the interest paid into the account. Similarly, there are three monthly expenditures, so the schematic should have three arrows pointing out of the box. To represent the expenses, let C denote the cell phone bill, R represent rent, and D represent the donation to the dog rescue. Finally, since the account was opened with a $4000 deposit, we'll make a note of that within the box by writing $A(0) = 4000$. Putting all of this together, we should have a schematic similar to the one displayed in Figure 4.1.

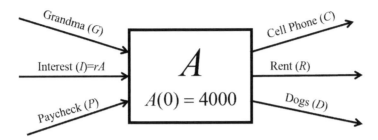

FIGURE 4.1: Schematic of Stephen's bank account.

Now that there is a schematic representing the changes in Stephen's bank account, we can derive a differential equation to model his account balance. We begin with a mathematical expression of the change in his account balance. This will be represented by the first derivative $\dfrac{dA}{dt}$. The change in account balance will be equal to the sum of the monthly deposits minus the monthly expenses. Using the schematic and the variables which were previously defined, the differential equation would be

$$\frac{dA}{dt} = G + I + P - C - R - D; \qquad A(0) = 4000$$

Note that this is a very generic equation, as there is no specific information about any of the deposits or expenses. They could all be constant values, time-dependent functions, or expressions that are dependent on the state variable A. As we get more information about each term, we can update the model to reflect the changes. For example, we were told that the monthly interest is proportional to the account balance. This means that $I = rA$, and the differential equation would become

$$\frac{dA}{dt} = G + rA + P - C - R - D; \qquad A(0) = 4000$$

■

Solving a First Order Differential Equation Model

Once a potentially suitable model has been constructed (or otherwise selected), our efforts must then turn to solving the differential equation. In general, there are two options for finding the solution. Using techniques learned in an introductory differential equations course or a computer algebra system, we can solve the differential equation analytically. Alternatively, we can use one of several algorithms to solve the equation numerically. There are benefits and drawbacks for both analytical and numerical solutions.

Analytical solutions are exact, and can be algebraically manipulated to answer a variety of questions. Numerical solutions are approximate and subject to some level of error. Both solutions can be used to produce a graph of the modeled state variable. Making changes to the model parameter tends to be easier with an analytical solution, where we can just change the parameter value within the solution. To make a similar change to a numerical solution, we would have to run the simulation again with the new parameter value. It should be pointed out that not all equations can be solved analytically, so a numerical solution is sometimes the only option. In the following examples, we will construct and solve models using analytical and numerical methods.

Example 4.2.2 *Dr. Marlin Rowe is a marine biologist who has been placed in charge of monitoring and predicting the fish population in the local lakes. He has hired you to assist him with the project for Lake Whosagudog by creating a mathematical model to predict the future population in the lake. After studying the lake, Dr. Rowe has determined that there are currently 3000 fish in the lake. Every month, 200 new fish are hatched, and 5% of the population dies. Using this information, you have been asked to find a formula to predict the number of fish in the lake after t months.*

We'll start by creating a model that incorporates the assumptions presented by Dr. Rowe. Since we want to model the fish population, we'll let our state variable be F. The current number of fish in the lake is 3000, so we'll have $F(0) = 3000$. To construct the differential equation, we have to consider the quantities that are entering and exiting the population. Fish enter the population through birth, and exit the population through death. We have been informed that the number of births per month is a constant 200, and the number of deaths per month is equal to 5% of the population at that time. We can assimilate all of this information into a schematic that will represent the dynamics of our model (see Figure 4.2).

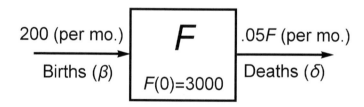

FIGURE 4.2: Schematic for the dynamics of the population of fish in Lake Whosagudog.

From our schematic, we can form an initial value problem that will represent the change in the fish population for Lake Whosagudog. The change in the population will be represented by $\dfrac{dF}{dt}$. We'll add the birth rate, which is 200 fish per month, and subtract the death rate, which is $0.05F$. This will give the following initial value problem.

$$\frac{dF}{dt} = 200 - 0.05F; \qquad F(0) = 3000 \tag{4.1}$$

This is a first order linear equation that can be solved analytically by using an integrating factor or by using separation of variables. We'll use the integrating factor here. We first need to rewrite the equation in standard form,

$$\frac{dF}{dt} + 0.05F = 200$$

Now that the equation is in standard form, we can identify $P(t) = 0.05$. Thus, the integrating factor for this equation will be $\mu(t) = e^{\int 0.05\,dt} = e^{0.05t}$. We'll multiply the equation by the integrating factor, which yields

$$e^{0.05t}\frac{dF}{dt} + 0.05e^{0.05t}F = 200e^{0.05t}$$

Using the product rule from calculus, we can rewrite the left hand side of the equation,

$$\frac{d}{dt}\left[e^{0.05t}F\right] = 200e^{0.05t}$$

Now, we integrate both sides and solve for our state variable, F.

$$\int \frac{d}{dt}\left[e^{0.05t}F\right]\,dt = \int 200e^{0.05t}\,dt$$
$$e^{0.05t}F = 4000e^{0.05t} + C$$
$$F = 4000 + Ce^{-0.05t}$$

Finally, we'll apply the initial condition $F(0) = 3000$ to solve for the arbitrary constant C

$$F(0) = 3000 \Rightarrow 3000 = 4000 + Ce^{-0.05(0)}$$
$$\Rightarrow 3000 = 4000 + C$$
$$\Rightarrow -1000 = C$$

So, the population of fish in Lake Whosagudog can be represented by the function $F(t) = 4000 - 1000e^{-0.05t}$. ∎

In the previous example, we found an analytical solution for the population of fish in a lake by using techniques from an introductory differential equations course. Alternatively, we could have found the analytical solution by employing a computer algebra system (CAS) such as *Wolfram Mathematica*. To find the analytical solution to a differential equation using *Mathematica*, we would use the *DSolve* command. The following lines represent how we would input the differential equation from the previous example into *Mathematica*, and its corresponding output. Please note that *Mathematica* is case sensitive and commands must be entered precisely (a capital s in the DSolve command, for example) in order to work.

input : DSolve[{F'[t]==200-0.05*F[t],F[0]==3000,$F[t]$,t]}

When the command is evaluated, it should generate the following output,

$$\text{output} : \left\{\left\{F[t] \rightarrow e^{-0.05t}\left(-1000. + 4000.e^{0.05t}\right)\right\}\right\}$$

Having an analytical solution to a differential equation gives us flexibility in the types of questions that we can answer regarding the model. With the analytical solution, we can produce a graph that represents the fish population over a period of time (see Figure 4.3), or we can use algebra to answer questions regarding time (when will the population reach a certain size?) or the size of the population.

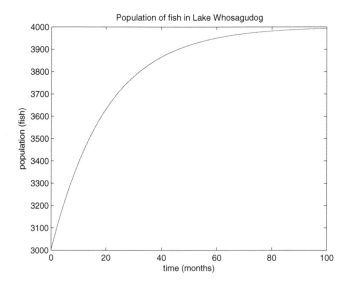

FIGURE 4.3: Simulated population of fish in Lake Whosagudog over a 100 month period.

Now, we will turn our attention to finding a numerical solution to a first order differential equation model.

Example 4.2.3 *During a routine maintenance visit, Phil the plumber discovers that your cylindrical water heater has developed a small hole and is leaking. As a precaution, he turns off the water leading to the heater, and tells you that he'll have to wait until all of the water inside leaks out before he can replace it. If the water heater is 6 feet tall and has a radius of 3 feet, and the hole in the heater is circular with radius $\frac{1}{4}$ inch, create a model for the height of the water within the tank. Graph the water level within the tank over the next 90 minutes.*

To determine the level of the water in the water heater at any given time, we first need to find a suitable differential equation. Note that it is not always necessary to construct a differential equation from scratch. There are many situations where the dynamics of what you are trying to model are well known, and equations already exist. For the leaky water heater, we can use a principle from hydrodynamics known as Torricelli's Law. Torricelli's Law states that for water that is draining through a small round hole, when the surface of the water is at a height h the water drains with the velocity that it would have if it fell freely from a height h. Equating the formula for kinetic energy ($\frac{1}{2}mv^2$) to the formula for potential energy (mgh) and solving for v would imply that the velocity of the water falling from height h would be equal to $\sqrt{2gh}$. Furthermore, the volume of the water draining through the hole (per second) can be found by multiplying the surface area of the hole A_H by the velocity of the draining water. This leads to the following differential equation for the change in velocity in the water heater.

$$\frac{dV}{dt} = -A_H\sqrt{2gh} \qquad (4.2)$$

Since our water heater takes the form of a right circular cylinder, the surface area of the water within will be a constant A_T, and the volume of water witin the tank can be written as $V = A_T h$. So, Equation (4.2) can be rewritten as

$$A_T\frac{dh}{dt} = -A_H\sqrt{2gh},$$

which leads us to the following differential equation representing the change in the height of the water within the tank.

$$\frac{dh}{dt} - -\frac{A_H}{A_T}\sqrt{2gh}$$

Now that we have a differential equation to model the height of the water in our leaky water heater, we can go about finding a numerical solution to the equation. In this example, we'll discuss finding a numerical solution using both *Mathematica* and *MATLAB*. We have the following parameter values that will need to be used to find the solution. $A_T = \pi \cdot (3)^2$ sq. ft., $A_H = \pi \cdot \left(\frac{1}{4}\right)^2$ sq. ft., and $g = 32\frac{\text{ft}}{\text{sec}^2}$.

To find the numerical solution for a differential equation using *Mathematica*, we'll use the *NDSolve* command. The following lines represent how we would input the differential equation for the height of the water in our tank into *Mathematica*, solve it, and generate a graph.

In1 : s = NDSolve[{$y'[t] == -(\text{pi} * (1/4)^2)/(\text{pi} * 3^2) * \text{Sqrt}[2 * (32) * y[t]]$,
$y[0] == 6$},y, {t, 0, 90}]

The above *Mathematica* command will generate the following output,

Out1 : {{$y \rightarrow$ InterpolatingFunction [Domain:{(0.,88.2)} Output:scalar]}} ,

which can be plotted using the following command:

In2 : Plot[Evaluate[y[t]/.s],{t,0,90},PlotRange $\rightarrow$ All, AxesLabel $\rightarrow$
{"time (min)", "height (ft)"}]

When *Mathematica* finds a numerical solution to the differential equation using the *NDSolve* command, it creates a vector of values that approximate the solution, which can be interpolated to represent the solution to the differential equation. The *Plot* command listed above instructs *Mathematica* to plot the interpolating function that was found when the differential equation was solved for time values between 0 and 90 minutes. The resulting graph is depicted in Figure 4.4(a).

We could also use *MATLAB* to find a numerical solution to our differential equation. To solve a differential equation using *MATLAB*, we need to create a function file that defines the differential equation. We can then solve the differential equation using one of the built in ODE solvers (such as *ode15*, *ode23*, or *ode45*). The following *MATLAB* code can be used to solve the differential equation that models our water heater and plot the results.

```
*****************************************************************
%Initialize time vector and Initial Condition(s)
tspan=0:.1:90;  %Define length of time (0 to 90 minutes)
initH=6;        %Define Initial height of water

%Call ODE solver
[t,h]=ode45(@WaterHeaterODE,tspan,initH);

%plot results
plot(t,h);
xlabel('time (min)');
ylabel('height (ft)');
%-----------------------------------------------------------------
```

```
function dhdt=WaterHeaterODE(~,h)
%Define parameters
AH=pi*(1/4)^2; %Area of the whole
AT=pi*3^2;   %Surface Area of the water in the tank
g=32;        %acceleration due to gravity (in ft/sec^2)

%Define ODE
dhdt=-AH/AT*sqrt(2*g*h);
end
```
****

When the MATLAB code above is run, it generates a graph that represents the height of the water in the water heater over time. The resulting graph is displayed in Figure 4.4(b).

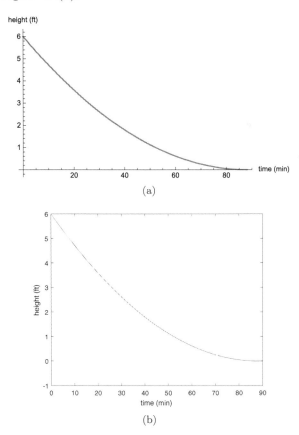

FIGURE 4.4: Height of the water in a leaking 6 foot water heater with respect to time, as computed numerically by (a)*Mathematica* and (b)*MATLAB*.

Exercises

1. Consider the model of Stephen's bank account constructed in Example 4.2.1. Assume that Stephen gets $50 per month from his grandmother, $800 a month from his paycheck, and the account has a 4.5% interest rate. Also, his cell phone bill is $100 per month, his rent is $550 per month, and he donates $35 dollars per month to the dog rescue. As stated in the example, the account was opened with $4000.

   (a) Write the differential equation that models his account under the given assumptions.

   (b) Find an analytical solution that represents the amount of money in Stephen's account at any given time (t).

   (c) ^{TE} Create a graph that represents the amount of money in the account over time.

   (d) Stephen would like to buy a new car, which will cost $12,000. How many months will it take for his account to reach that value?

2. During a winter snowstorm, the town of Autumndale deploys its fleet of snowplows after 4 inches of snow have accumulated on the streets. Assume that the snow continues to fall at a rate of $\frac{3}{5}$ inch per hour, and that the snowplows can remove the snow at a rate that is equal to 0.45 times the amount of snow on the ground per hour. Using this information, complete the following.

   (a) Formulate an initial value problem that can be used to model the depth of the snow on the ground in Autumndale.

   (b) Find an analytical solution to the initial value problem.

   (c) ^{TE} Plot the solution that represents the depth of the snow on the ground for an appropriate amount of time.

   (d) Is it possible under these assumptions for the fleet of snowplows to clear the snow from the road while it is still snowing? If so, when will the streets be clear of snow?

3. ^{TE} Pharmacokinetics is the study of how drugs are transported throughout the body. The most simple pharmacokinetic model treats the entire body as one compartment. Consider a hospital patient being treated for a rare

tropical disease with Professor Utonium's Chemical X. Chemical X is administered intravenously to the bloodstream at a rate of α milligrams per liter per hour, and is cleared from the bloodstream (via metabolism and excretion) at a rate that is proportional to the concentration within the bloodstream (assume a proportionality constant of k).

(a) Formulate a generic differential equation to represent the change in the concentration of Chemical X in the bloodstream over time.

(b) Assume that the drug is administered at a constant rate (25 milligrams per liter per hour) and is removed with a proportionality constant of 0.35 per hour. Find a solution that will represent the concentration of Chemical X within the bloodstream if the patient started with no Chemical X in his system to begin with.

(c) Assume that the patient is given one initial dose of 400 milligrams per liter, and no further drug is administered. Find a solution that will represent the concentration of Chemical X within the bloodstream in this scenario.

(d) If the doctor wants to assure that at least 60 milligrams per liter are in the patient's system at 10 hours, which dosing strategy should be employed?

4. Kirchhoff's second law states that, for a series circuit containing an inductor and a resistor, the sum of the voltage drop across the inductor and the voltage drop across the resistor is the same as the voltage applied to the circuit. This is represented by the differential equation

$$L\frac{di}{dt} + Ri = V(t),$$

where L is the inductance, R is the resistance, $i(t)$ is the current, and $V(t)$ is the voltage applied to the circuit. Consider an LR series circuit whose inductance is $\frac{1}{4}$ henry, and whose resistance is 5 ohms. Assume that the circuit is connected to a 9-volt battery and that the initial current is zero amperes.

(a) Find an analytical solution that represents the current $i(t)$ at any given time t.

(b) ^{TE} Graph the current across the circuit over time.

(c) What happens to the current in this series circuit as $t \to \infty$?

5. ^{TE} The schematic depicted in Figure 4.5 represents the dynamics of a population of trout in a local pond, where T represents the number of trout (in thousands).

   (a) Using the schematic, create the differential equation that would be used to model this trout population.

   (b) Assuming that the birth rate constant (α) is 2 per month, the restocking rate (λ) is 50 trout per month, the rate at which trout are caught (or harvested) is 35 trout per month, and the death rate constant is .75 per trout per month, find and plot the numerical solution for the population of trout over the next 5 months.

   (c) Does your solution suggest that the population is increasing or decreasing over time? Which parameter value (or values) would you need to modify in order to change the outcome of the population over time?

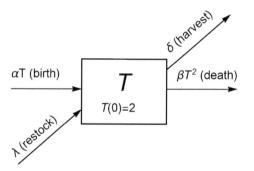

FIGURE 4.5: Schematic of a population of trout.

6. ^{TE} One method of modeling the growth of a tumor is with the Gompertz Equation, which takes the form

$$\frac{dP}{dt} = P(a - b \ln P),$$

where P is the population of cells (in hundreds). Consider a tumor with an initial population of 500 cells, $P(0) = 5$.

   (a) Plot the numerical solution for the growth of the tumor using the Gompertz Differential Equation with $a = 2$ and $b = 1$.

   (b) Plot the numerical solution for the growth of the tumor using the Gompertz Differential Equation with $a = 2$ and $b = 3$.

(c) Plot the numerical solution for the growth of the tumor using the Gompertz Differential Equation with $a = 3$ and $b = 1$.

(d) Using the results from above, describe how the change in the parameters a and b affects the predicted population over time.

Project^{TE}: Cleaning Up a Polluted Lake

You have been hired by the city of Autumndale to help monitor and clean up Lake Autumndale, the centerpiece of Obediah Autumndale Municipal Park. It seems that B. Tuscaloosa Scalds, CEO of Autumndale Nuclear Power, has been dumping a chemical pollutant into the Twisty River, which feeds into Lake Autumndale from the melting snow of Mount Useful. Water is released from the polluted lake via the Autumndale Dam at the same rate at which it enters the lake from the Twisty River. When you arrive at the scene there is 150 pounds of pollutant in the 300,000 gallon lake. Use your knowledge of differential equations to examine the following problems.

1. **Let bygones be bygones...** Initially, the city of Autumndale doesn't want to take any action about the lake. Mr. Scalds continues to dump enough pollutant into the Twisty River so that there is 1 pound of pollutant in every 50 gallons of water. Assume that the Twisty River flows into Autumndale Lake at a rate of 700 gallons per day, and the Autumndale Dam releases water at the same rate.

   (a) Form an initial value problem that can be used to model the amount of pollutant in the lake at any given time.

   (b) Find the solution to the inital value problem that you formed in part a.

   (c) Plot the solution to the initial value problem.

   (d) Describe what happens to the level of pollutant in the lake as time goes by.

2. **Let nature heal itself...** Another option that the city of Autumndale is considering is to force Mr. Scalds to stop dumping pollutant into the Twisty River, and have the pollution get filtered out naturally. Under this option, there would be no pollutant entering the lake via the Twisty River. Pollutant would be removed from the lake when the Autumndale Dam releases water at the rate of 700 gallons per day.

(a) Form an initial value problem that can be used to model the amount of pollutant in the lake at any give time under this scenario.

(b) Find the solution to the initial value problem that you formed in part a.

(c) Plot the solution to the initial value problem.

(d) Describe what happens to the level of pollutant in the lake as time goes by.

(e) The Autumndale Chamber of Commerce wants to hold a swimming competition at Lake Autumndale in 150 days, but in order for the event to take place, the amount of pollutant in the lake must be less than 10 pounds. From your results, will they be able to hold their swim meet?

3. **We're here to pump you up...** For a third option, in addition to forcing Mr. Scalds to stop dumping pollutant (pollutant is no longer entering the lake), you have been authorized to install a series of pumps into Lake Autumndale. Your pumps are capable of removing a certain amount of pollutant from the lake per day without affecting the volume of the lake. So, pollutant will be removed from the lake when the Autumndale Dam releases water at the rate of 700 gallons per day, as well as through your pump system.

(a) Assume that your pumping system is capable of removing an additional 1 pound of pollutant per day. Form an initial value problem that can be used to model the amount of pollutant in the lake at any given time under this scenario.

(b) Find the solution to the initial value problem that you formed in part a.

(c) Plot the solution to the initial value problem.

(d) Describe what happens to the level of pollutant in the lake as time goes by.

(e) The Autumndale Chamber of Commerce wants to hold a swimming competition at Lake Autumndale in 150 days, but in order for the event to take place, the amount of pollutant in the lake must be less than 10 pounds. From your results, will they be able to hold their swim meet? When, if ever, will the lake be free of pollutant?

Project^{TE}: Crime Scene Investigation: Differential Equations Unit

Congratulations, you have been selected for an internship with the local police department's forensics team. Your first case involves a dead body that was found submerged in a small pond. You have been asked to use Newton's Law of Cooling/Warming to help model the temperature of the body after its death. Newton's Law of Cooling is modeled with the following differential equation,

$$\frac{dT}{dt} = k(T - T_m),$$

where $T(t)$ represents the temperature at time t, k represents the rate constant at which the body temperature increases or decreases (per minute), and T_m represents the surrounding temperature of the body (°F). Assume that for the body that was found in the pond, k is -0.0045, and at the time of death, the body temperature was 98.8°F. The detectives investigating the case have several different theories of what happened, and they are hoping that your modeling efforts can help them determine the true events.

1. **This is a nice location.** The first theory is that the victim was fishing with Mr. Forest at the pond when he was murdered. The body has been submerged in the 62°F water for the past 4 hours.

   (a) Solve the differential equation to get an expression for the temperature of the body since its time of death.

   (b) Plot the solution of the differential equation over the 4 hour time period.

   (c) Describe what is happening to the body temperature over time in this scenario.

2. **A body on the move.** The second theory is that the victim got into an argument with Professor Eggplant while helping him restock his ice cream cooler. The argument led to the victim's death. Not knowing what to do with the body, Professor Eggplant left it in the -20°F freezer for 2 hours, and then dumped it into the 62°F pond, where it was discovered 3 hours later.

   (a) Solve the differential equation to get an expression for the temperature of the body since its time of death. Note that your solution will take the form of a piecewise function.

   (b) Plot the solution of the differential equation over the 4 and a half hour time period.

  (c) Describe what is happening to the body temperature over time in this
      scenario.

3. **The well-travelled body.** The last theory is that the victim was gar-
   dening in the greenhouse with Colonel Goldenrod when he met his demise.
   Colonel Goldenrod left the body in the 105°F greenhouse for 1 hour. He
   then moved the body to the meat cooler at his family's steakhouse, where
   the temperature was 45°F. The body stayed in the cooler for 2 hours. With
   an impending visit from the health inspector, Colonel Goldenrod had to
   move the body once again. He dumped it in the pond (62°F), where it was
   discovered 2.5 hours later.

   (a) Solve the differential equation to get an expression for the temperature
       of the body since its time of death. Note that your solution will take
       the form of a piecewise function.

   (b) Plot the solution of the differential equation over the 5 and a half hour
       time period.

   (c) Describe what is happening to the body temperature over time in this
       scenario.

4. **Here is what really happened.** The detectives have informed you that
   the temperature of the body when it was discovered was approximately
   56°F. Given this information and your results from the three scenarios,
   which scenario is most likely? Who should be arrested for murder?

4.3 Systems of First Order Differential Equations

Often when creating mathematical models, we wish to examine the change in multiple, related quantities with respect to time. To accomplish this, we would utilize a system of differential equations. If we consider multiple quantities or states that are simultaneously changing and potentially affecting each other over time, a system of equations can be constructed and solved where each equation models one of the states. These systems are often called *coupled systems* because the derivative of each state variable is dependent on the other state variables.

Constructing a System of First Order Differential Equations

To construct a system of first order differential equations, we'll again draw a schematic to represent the interactions of the different modeled populations. We will represent the state variables with boxes and the interactions with arrows. From the schematic, we can formulate the system of differential equations. There will be an equation for each state variable (or box). The derivative of the state variables will be equal to the changes in the population. Interactions entering the population will be added and interactions causing a loss in the population will be subtracted.

Example 4.3.1 *Officer Crumbcake from the Eastside Police department is in charge of the anti-gang task force. He wants to create a mathematical model to keep track of the populations of the two rival gangs in the city (the Dolphins and the Planes). Both gangs recruit members from the general population, and from the other gang. Members are removed from the gang by returning to the general population, joining the other gang, or through death by interactions between the two gangs. Using these dynamics, draw a schematic to represent the Dolphins, the Planes, and the General Population. Use your schematic to establish a system of differential equations that could be used to model the population of each group.*

To sketch the schematic representing the populations of interest, we'll begin by drawing three boxes to represent our three state variables. We'll let D represent the Dolphins gang, P represent the Planes gang, and G represent the general population. We'll assume that people join the Dolphins and the Planes from the general population at rates that are propotional to the general population. Let α_D and α_P be the rate constants for joining the Dophins and Planes, respectively. We'll represent this interaction by drawing two arrows: one that leads from the general population (G) to the Dolphins (D) and one that leads from G to the Planes (P).

Additionally, it is possible for members of one gang to switch to the other gang. This will be represented by an arrow from D to P, and another from P to D. We'll assume that the defection of gang members occurs as a percentage of the gang's population, with rate constants of μ_P for those leaving the Planes for the Dolphins, and μ_D for those leaving the Dolphins for the Planes.

We'll represent members leaving the gang and returning to the general population with an arrow from each gang leading back to the general population. We'll assume that members leave the Dolphins and Planes at a rate that is proportional to the gang's population, with rate constants ω_D and ω_P, respectively.

Finally, gang members can be killed through interactions with the other gang. We represent the deaths from gang "wars" with an arrow leaving each gang popuation. The rate at which gang members are killed will be proportional to the product of the two gang populations. We'll use δ_D and δ_P to represent the rate constants for the gang-related deaths of members of the Dolphins and Planes, respectively. Note that this model is being constructed with an assumption that only gang members are killed. Often when modeling populations, it is assumed that the general birth rate and death rate are equal to each other, creating a closed population. The assumption of a closed population is true of this model, although not explicitly stated, with the addition of the potential deaths due to gang violence.

This will complete the schematic for the interactions of the three populations as outlined by the problem, and the schematic is presented in Figure 4.6.

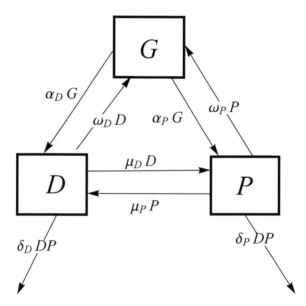

FIGURE 4.6: Schematic for the interactions of two rival gangs, the Dolphins (D) and the Planes (P), with the general population G of a city.

Using the schematic, we can formulate the system of differential equations that will represent the dynamics of the gang populations. Each state variable (represented by the boxes in the schematic) will have its own differential equation. The change in the state variable (its derivative) will be set equal to the sum of the amounts entering the box minus each component leaving the box. This will lead to the following system of differential equations.

$$\frac{dG}{dt} = \omega_D D + \omega_P P - \alpha_D G - \alpha_P G \tag{4.3}$$

$$\frac{dD}{dt} = \alpha_D G + \mu_P P - \omega_D D - \mu_D D - \delta_D DP \tag{4.4}$$

$$\frac{dP}{dt} = \alpha_P G + \mu_D D - \omega_P P - \mu_P P - \delta_P DP \tag{4.5}$$

So, using the schematic of the interactions of the gangs with the general population, we have constructed a system of differential equations which can be used to model the populations over time. Note that in order to solve this system, we would need to specify the initial populations and values for the various rate constants. The results generated by the model will vary based on the parameterization.

■

Solving a System of First Order Differential Equations

Again, we have two options for solving a system of differential equations. We can find an analytical solution by solving the system by hand or by using a computer algebra system. We can also find a numerical solution to the system using an appropriate algorithm in a computational system. Here, we will work an example of each.

Example 4.3.2 *After watching Willy Wonka and the Chocolate Factory late one night, the CEO of Happy Cow Dairy company has requested your assistance with a project. In the movie, the Willy Wonka Chocolate Factory mixes their chocolate by waterfall. Happy Cow Dairy wishes to do the same with their chocolate milk. You have been asked to construct a mathematical model that can be used to simulate the mixing of the chocolate and the milk in their production facility. Two 5000 gallon reservoirs have been constructed which the milk will flow between. The milk will flow from the upper reservoir to the lower reservoir as a "waterfall" at a rate of 50 gallons per minute. Milk will then be pumped back into the upper reservoir from the lower reservoir at the same rate. Initally, 1000 pounds of chocolate is dissolved in the upper reservoir. Assuming that both reservoirs are well-mixed, construct a system of differential equations that can predict the amount of chocolate in each reservoir at any given time. Use your model to create a graph depicting the amount of chocolate in each reservoir over time.*

We'll start by drawing a schematic to represent the system. Let U represent the amount of chocolate (in pounds) in the upper reservoir and L represent the amount of chocolate in the lower reservoir. Draw two boxes to represent these state variables. Milk flows from U to L and back from L to U, so there should be an arrow between our two boxes going in each direction. To determine the amount of chocolate flowing in each direction, we'll need to multiply the flow rate by the concentration of chocolate in the reservoir from which the flow originated. The flow rate is known to be 50 gallons per minute, and since each reservoir is assumed to be well-stirred, the concentration within can be represented by the amount of chocolate (either U or L) divided by the volume. Putting this together, the amount of chocolate flowing from U to L will be $\frac{U}{100}$ pounds per minute and the amount flowing from L to U will be $\frac{L}{100}$ pounds per minute. The schematic for the waterfall mixing system is depicted in Figure 4.7. From the schematic, we construct the following system of differential equations

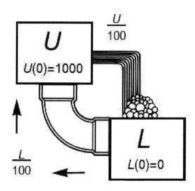

FIGURE 4.7: Schematic for the mixing of chocolate milk using and upper and lower reservoirs.

to model the mixing of the chocolate milk.

$$\frac{dU}{dt} = \frac{L}{100} - \frac{U}{100} \qquad (4.6)$$

$$\frac{dL}{dt} = \frac{U}{100} - \frac{L}{100} \qquad (4.7)$$

To find an analytical solution to the system, we'll first solve the system using methods from an introductory differential equations course. Begin by rewriting the system using differential operators, yielding

$$DU = \frac{L}{100} - \frac{U}{100}$$

$$DL = \frac{U}{100} - \frac{L}{100}.$$

Move all state variables to the left hand side of the system via addition/subtraction, and combine like terms to get

$$\left(D + \frac{1}{100}\right)U - \frac{1}{100}L = 0 \tag{4.8}$$

$$-\frac{1}{100}U + \left(D + \frac{1}{100}\right)L = 0. \tag{4.9}$$

We'll solve the system using the elimination method. We can eliminate U if we multiply (4.8) by $\frac{1}{100}$ and (4.9) by $\left(D + \frac{1}{100}\right)$ and add the resulting equations together. This yields

$$\left[\left(D + \frac{1}{100}\right)^2 - \frac{1}{1000}\right]L = 0,$$

which can be simplified to

$$\left(D^2 + \frac{1}{50}D\right)L = 0.$$

To solve for L, we form the auxiliary equation $m^2 + \frac{1}{50}m = 0$. The auxiliary equation has roots $m = 0$ and $m = -\frac{1}{50}$, which means the solution for L will be

$$L = c_1 + c_2 e^{-\frac{t}{50}}.$$

Now that we have a solution for L, we can compute a solution for U by using one of our original equations. We can solve (4.7) for U and insert the solution for L, yielding the following:

$$U = 100\left[\frac{dL}{dt} + \frac{L}{100}\right]$$

$$\Rightarrow \quad U = 100\left[\frac{-c_2 e^{-\frac{t}{50}}}{50} + \frac{c_1 + c_2 e^{-\frac{t}{50}}}{100}\right]$$

$$\Rightarrow \quad U = c_1 - c_2 e^{-\frac{t}{50}}.$$

Finally, we will use the given initial conditions to find the particular solutions to the system. We have $U(0) = 100$ and $L(0) = 0$ as our initial conditions. If we apply these to our solutions, we will get the following system:

$$U(0) = 100 \quad \Rightarrow \quad 100 = c_1 - c_2$$
$$L(0) = 0 \quad \Rightarrow \quad 0 = c_1 + c_2.$$

Solving the resulting system of equations, we find that $c_1 = 50$ and $c_2 = 50$, so the solutions to our system of equations will be

$$U(t) = 50 - 50e^{-\frac{1}{50}t}; \qquad L(t) = 50 + 50e^{-\frac{1}{50}t}.$$

We have analytically found the solution that models the mixture of chocolate between the two reservoirs. While we found our solution by hand, we also could have used a computer algebra system to find the solution to the system. We could solve the system using the following *Mathematica* command:

In1 : sol=DSolve[{$x'[t]$ == $1/100 * y[t] - 1/100 * x[t]$,$y'(t)$ == $1/100 * x[t]$
$-1/100 * y[t]$, $x[0]$ == 100, $y[0]$ == 0},{$x[t]$,$y[t]$},t]

The above *Mathematica* command generates the following output.

Out1 : $\left\{\left\{ x[t] \to 50e^{-t/50}\left(1+e^{t/50}\right), y[t] \to 50e^{-t/50}\left(-1+e^{t/50}\right)\right\}\right\}$

Furthermore, we can plot the solutions for the amount of chocolate in each reservoir over time by using the following command:

In2 : Plot[Evaluate[{$x[t]$,$y[t]$}/.sol],{t,0,120},PlotStyle→{Black,{Red,
Dashed}},AxesLabel→{"time(min)","Amount of Chocolate(lbs)"}]

The graph that is generated using this command is presented in Figure 4.8. ∎

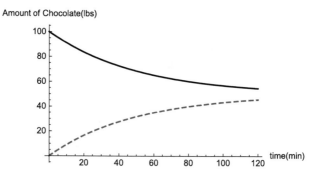

FIGURE 4.8: The amount of chocolate in the upper (solid line) and lower (dashed line) reservoirs over 120 minutes.

The previous example illustrated how an analytical solution to a system of first order differential equations can be found by hand, or with the aid of a computer algebra system. We will now turn our attention to finding a numerical solution to a system of first order equations.

Example 4.3.3 *Industrial Occupations University (IOU) is a local trade school that wants to grow its enrollment. Currently, it has 300 freshmen, 200 sopho-mores, 250 juniors, and 225 seniors enrolled. It has developed a new retention plan, in which it hopes that 85% of the freshman class moves to the sophomore level, 75% of sophomores become juniors, 68% of juniors become seniors, and 75% of seniors graduate each year. Each year, 5% of each class leaves the school by dropping out or transferring to another school. Additionally, IOU plans to recruit 75 students more than the number of graduating Seniors each year. Us-ing this information, formulate a system of equations to model the population of each class in the school. Create a graph that depicts the population of each class over the next 20 years and another figure that illustrates the total population over the next 20 years. Will their proposed plan increase the overall enrollment at the school?*

Again, we'll begin by drawing a schematic of the dynamics of the population within the school. We'll start by drawing boxes to represent our state variables. We have freshmen (F), sophomores (P), juniors (J), and seniors (S). We'll then draw arrows between the state variables to represent the interactions between them. Let α represent the rate at which freshmen become sophomores (%/year), β represent the rate at which sophomores become juniors (%/year), δ represent the rate at which juniors become seniors (%/year), and γ be the rate at which seniors graduate from the school (%/year). Also, we'll let μ denote the rate at which students leave the school from each class due to dropping out or transferring. We are told that 75 more students are recruited each year than are expected to graduate, so the students entering the freshmen class will be represented by the expression $75 + \gamma S$. Putting all of this together, we should arrive at a schematic similar to that depicted in Figure 4.9.

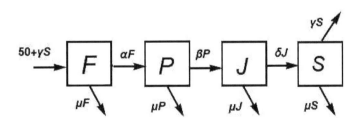

FIGURE 4.9: A schematic for the population of the classes at Industrial Occu-pations University.

Using the schematic in Figure 4.9, we can derive the following system of differential equations to model the population at the school.

$$\frac{dF}{dt} = 75 + \gamma S - \alpha F - \mu F; \qquad F(0) = 300 \qquad (4.10)$$

$$\frac{dP}{dt} = \alpha F - \mu P - \beta P; \qquad P(0) = 200 \qquad (4.11)$$

$$\frac{dJ}{dt} = \beta P - \mu J - \delta J; \qquad J(0) = 250 \qquad (4.12)$$

$$\frac{dS}{dt} = \delta J - \mu S - \gamma S; \qquad S(0) = 225 \qquad (4.13)$$

It may be beneficial to organize the information that was given to us in the retention plan. The parameter values that we'll be using to solve the system are reported in Table 4.1 We'll first solve the system of equations using *Math-*

TABLE 4.1: Parameter Descriptions and Values for the Population Model for IOU

Parameter	Description	Value (per year)
α	Retention rate from freshman to sophomore year	.85
β	Retention rate from sophomore to junior year	.75
δ	Retention rate from junior to senior year	.68
γ	Graduation rate for the senior class	.75
μ	Drop-out/transfer rate from each class	.05

ematica. To numerically solve the system, we'll use the following command in *Mathematica*:

In1 : sol2=NDSolve[{$f'[t] == 75 + .75 * s[t] - .85 * f[t] - .05 * f[t]$,
      $p'[t] == .85 * f[t] - .05 * p[t] - .75 * p[t]$,
      $j'[t] == .75 * p[t] - .05 * j[t] - .68 * j[t]$,
      $s'[t] == .68 * j[t] - .05 * s[t] - .75 * s[t], f[0] == 300, p[0] == 200$,
      $j[0] == 250, s[0] == 225\}, \{f[t], p[t], j[t], s[t]\}, \{t, 0, 20\}]$

Mathematica will display the creation of four interpolating functions as solutions to the system. To create a plot of the populations of the four classes, we'll use the following command:

In2 : Plot[Evaluate[{$f[t], p[t], j[t], s[t]$}/.sol2],{t,0,20},PlotStyle →
      {Black,Red,{Black,Dashed},{Red,DotDashed}},
      AxesLabel → { "time (years)", "Class Population"}]

Furthermore, if we want to create a graph that represents the total population of the school over the years, we can use the following command:

In3 : Plot[Evaluate[$(f[t] + p[t] + j[t] + s[t])$/.sol2],$\{t,0,20\}$,
PlotStyle → {Black},AxesLabel → { "time (years)",
"Total Population" }]

The resulting graphs from these *Mathematica* commands are shown in Figure 4.10 We'll also numerically solve the system of differential equations using

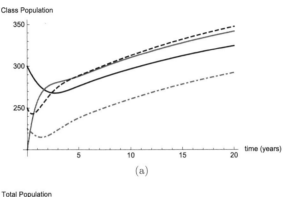

(a)

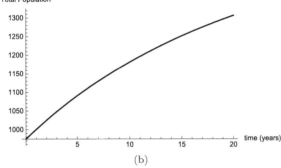

(b)

FIGURE 4.10: (a) Class populations (freshmen (solid black), sophomores (solid red), juniors (black dashed), and seniors (red dot-dashed)) and (b) Total population of IOU as simulated by *Mathematica*.

MATLAB. Again, to solve differential equations in *MATLAB*, we'll need to define the system within a function, and then use one of the pre-defined solvers to create the solution to the system. The following code was used to solve the system of equations and produce the graphs depicted in Figure 4.11.

```
*****************************************
%Initialize time vector and Initial Condition(s)
tspan=0:.01:20;  %Define length of time (0 to 20 years)
```

```
%Initial Conditions
F0=300;   %Initial Freshmen
P0=200;   %Initial Sophomore
J0=250;   %Initial Juniors
S0=225;   %Initial Seniors

IC=[F0;P0;J0;S0];
%Call ODE solver
[t,y]=ode45(@SchoolPopODE,tspan,IC);
F=y(:,1); P=y(:,2); J=y(:,3); S=y(:,4);
%plot class results
figure
plot(t,F,'k',t,P,'r',t,J,'k--',t,S,'r-.');
xlabel('time (years)');
ylabel('Class Populations');
legend('Fresh','Soph','Junior','Senior')
%plot total population
figure
plot(t,F+P+J+S,'k')
xlabel('time (years)');
ylabel('Total Populations');
%---------------------------------------------------------------------

function dx=SchoolPopODE(~,x)
%Define parameters
alpha=.85; %Retention from F to P
beta=.75;  %Retention from P to J
delta=.68; %Retention from J to S
gamma=.75; %Graduation of Seniors
mu=.05; %loss rate due to dropouts/transfers
%Define States
F=x(1); P=x(2); J=x(3); S=x(4);
%Define ODES
dx=[75+gamma*S-alpha*F-mu*F;
    alpha*F-mu*P-beta*P;
    beta*P-mu*P-delta*J;
    delta*J-mu*S-gamma*S];
end
**************************************************
```

Examination of the results displayed in Figures 4.10 and 4.11 created with *Mathematica* and *MATLAB*, respectively, shows that under the proposed retention and recruitment plan, the population of IOU should increase by approximately 350 students over the next 20 years.

■

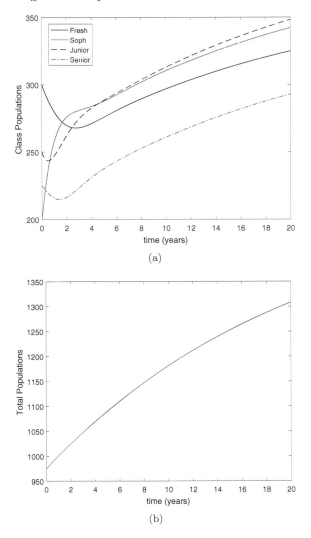

FIGURE 4.11: (a) Class populations (freshmen (solid black), sophomores (solid red), juniors (black dashed) and seniors (red dot-dashed)) and (b) Total population of IOU as simulated by *MATLAB*.

Exercises

1. ^{TE} Pepsilon and Calcu-Cola have both developed a new flavor which they will sell for a month in direct competition with each other. Pepsilon is hoping that its chili-cheese cola will be more popular than Calcu-Cola's smores soda. Market analysts have determined that the sales of the new

beverages can be modeled with the following competition model.

$$\frac{dP}{dt} = P(1 - 0.3P - 0.4C)$$

$$\frac{dC}{dt} = C(0.5 - 0.1 * C - 0.4P),$$

where P and C are the sales (in thousands) of the new Pepsilon and Calcu-Cola drinks, respectively.

(a) Assume that Pepsilon has initial sales of 2000 ($P(0) = 2$), and Calcu-Cola has initial sales of 5000 ($C(0) = 5$). Plot the solutions to the system of equations over 30 days. What happens by the end of the month?

(b) Assume that Pepsilon's initial sales were 5000, and Calcu-Cola initially sold 2000. Plot the sales for each company over the 30 day period. Describe the sales for each company.

(c) Consider the scenario where each company sold 5000 units initially. Plot the sales for each company over the 30 day period. Which company fares better over the promotional period?

2. Shiny Happy Medicine Company is developing a new drug which it will call Cure-All. They are in the process of testing the drug in clinical trials. Currently, they are adding the drug to a solution in which a cell is suspended and observing how the cell absorbs the drug. Fick's Law of diffusion states that the rate at which a chemical passes through a cell membrane is proportional to the concentration gradient (the difference between the concentration inside and outside of the cell). We can construct a model for the concentrations of Cure-All in the solution and within the cell using Fick's Law, as follows:

$$\frac{dS}{dt} = \gamma(C - S)$$

$$\frac{dC}{dt} = \gamma(S - C),$$

where S is the concentration of Cure-All in the solution (μM), C is the concentration of Cure-All within the cell (μM), and γ is the diffusivity rate constant (1/sec) of the cell membrane. Assume that the researchers have placed a cell into a solution with 50 μM of Cure-All, and that the diffusivity rate constant (γ) is equal to 0.675.

(a) Find the analytical solution to the system of equations.

(b) Using your analytical solution, determine when, if ever, the concentration within the cell will be the same as that in the solution.

(c) ^{TE} Plot the graph of the solutions to verify your answer from above.

3. ^{TE} Sean Lassiter describes himself as a history buff. He is particularly interested in the Civil War. As a project in his mathematical modeling class, he constructed the following system of differential equations to model the Battle of Gettysburg.

$$\frac{dU}{dt} = -\alpha U C$$
$$\frac{dC}{dt} = -\beta U C,$$

where U represents the number of Union soldiers, C represents the number of Confederate soldiers, and α and β are the rate constants for the casualty rate for the Union and Confederacy, respectively. Sean found some information online [24]. The battle involved 82,289 Union soldiers and 75,000 Confederate soldiers. Over the 3 day battle, the Union suffered 23,000 casualties and the Confederacy amassed 28,000 casualties.

(a) In his first attempt at creating the model, Sean assumed α to be 1×10^{-6} and β to be 1×10^{-7}. Solve the system of equations with these assumptions. Plot the graph of the solutions, and explain why Sean knew that he had chosen inappropriate parameter values.

(b) After realizing his error, Sean assumed $\alpha = 8 \times 10^{-8}$, and $\beta = 1 \times 10^{-7}$. Plot the solution to the system using these parameter values. Are these results more reasonable than the first attempt?

4. ^{TE} Ranger Jones is in charge of all scientific studies at the world reknowned Jellystone national park. Currently, he is observing the relationship between the populations of rabbits and wolves within the park boundaries. Wolves are natural predators of rabbits. Figure 4.12 depicts a schematic for the dynamics of the population of rabbits (R) and wolves (W) within the park.

(a) Use the schematic in Figure 4.12 to derive a system of equations to model the populations of rabbits and wolves within the park.

(b) The model illustrated in the schematic and represented by the equations that you have written is known as the Lotka-Volterra predator-prey model. Describe what each term in your model might physically

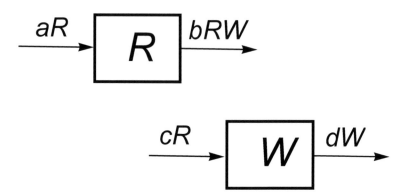

FIGURE 4.12: Schematic of rabbit and wolf population dynamics.

represent.

(c) ^{TE} Ranger Jones has determined that there were initally 200 rabbits in the park and 5 wolves. Assume that $a = 5, b = .75, c = 0.3$, and $d = 0.015$. Find and plot the solutions to the system of equations over the course of 36 months. Describe what happens to the rabbit and wolf populations with respect to each other.

5. ^{TE} Buck Smith plans to open a local hunting reserve and would like to create a mathematical model to predict the population of deer on the property. The deer will be split into three groups: youth (Y), adult(A), and senior(S). Deer are born into the youth population at a rate that is $1/4$ of the adult population per year. Every year, $1/3$ of the youth population grow into adults. The adult population is the only one which is allowed to be hunted, and the reserve plans to allow 200 to be killed per year. Each year, $1/10$ of the adult population grows into seniors. Senior deer are protected from hunting, but they die of natural causes at a rate that is $1/5$ of their population per year.

(a) Given this information, draw a schematic to represent the dynamics of the deer population on this huning reserve.

(b) Using your schematic, create a system of differential equations that could be used to model the populations of young, adult, and senior deer within the reserve.

(c) Assume that there were initially 200 young deer, 1000 adult deer, and 300 senior deer in the reserve. Solve your system of equations, plot the results, and describe what is happening to the deer population over the next 30 years. Will the hunting reserve be able to stay in business?

(d) In an effort to ensure a successful business, Buck reduces the number of yearly kills from 200 to 100. Modify your model to reflect this change, and plot the solution. Will the population survive with this modification?

6. The common cold is a communicable disease that can be contracted through contact with an infected person. One annoying aspect of the common cold is that after recovering from a cold, people can catch a cold again. A model of the spread of a common cold can be constructed using an SIS model, where the population is divided into susceptible (S) individuals and infected (I) individuals. Susceptible individuals become infected at a rate of $\alpha SI/N$ people per day, where α is the infection rate. Infected individuals return to the susceptible population at a rate of γI people per day, where γ is the recovery rate.

(a) Draw a schematic to represent the dynamics of the SIS model.

(b) Using the schematic, create a system of differential equations for the SIS model.

(c) Assume that the common cold has an infection rate of 50% per day, and a recovery rate of 1/5 per day. Use your model to predict the spread of the common cold over 60 days in a population of 50050 with 50040 susceptible people and 10 infected people initially.

Project^{TE}: Why Would You Jump from a Perfectly Good Plane?

During your summer vacation, you have been hired by the What Was I Thinking© Extreme Sports Company as a consultant. This year, they want to offer skydiving adventures to their clients, but they need your assistance in planning the details of the skydive. They plan to fly their plane at 8000 ft, where their customers will jump from the plane with an initial velocity of 0 ft/sec. After a brief period of free fall, the customer is to pull their ripcord, deploying their parachute. The velocity and position of a falling mass is modeled using the following system of differential equations,

$$\frac{ds}{dt} = v$$

$$m\frac{dv}{dt} = -mg + kv^2,$$

where $s(t)$ represents the position above the ground in feet, $v(t)$ represents the velocity of the skydiver in feet per second, m is the mass of the skydiver in kilograms, g is the acceleration due to gravity in feet per second squared, and k is the rate coefficient of the air resistance in one per second. For all skydivers, g will be 32 $\frac{\text{ft}}{\text{sec}^2}$, and k will be equal to 0.75 $\frac{1}{\text{sec}}$ while the diver is in free fall.

1. **The average customer.** The average customer with all of the skydiving gear has a mass of 85 kg. For the typical parachute used by the company, $k = 12$. The owners of the company are instructing their customers to pull their ripcord 30 seconds after jumping from the plane.

   (a) Solve the system of differential equations to show the velocity and height of the average skydiver over time.

   (b) How long will it take for the skydiver to reach the ground after jumping from the plane?

   (c) In order to have a safe landing without sustaining any injuries, the velocity of the jumper must be less than 20 feet per second. Under the average customer scenario, would the customer have a safe landing by deploying their parachute after 30 seconds?

2. **You're going to need a bigger chute.** The company wants to be able to cater to larger individuals as well. Assume that you have a customer who, with all of the gear, has a mass of 120 kg. For individuals whose mass is greater than 100 kg, company policy is to use a bigger, stronger parachute with a k value of 20. They will still pull the ripcord to their parachute after 30 seconds.

   (a) Solve the system of differential equations to show the velocity and height of the larger (120 kg) skydiver over time.

   (b) How does the descent of the larger customer compare to that of the average customer?

   (c) How long will it take for the skydiver to reach the ground after jumping from the plane?

   (d) In order to have a safe landing without sustaining any injuries, the velocity of the jumper must be less than 20 feet per second. Under the average customer scenario, would the customer have a safe landing by deploying their parachute after 30 seconds?

3. **I've got the need for speed.** Some of the customers of the company are extreme thrill seekers. They want to wait until the last possible second to pull their ripcord. As a safety precaution, the company tells customers that they must pull their ripcord when they are 1000 feet from the ground. What is the latest time that an average person could safely pull the ripcord? What is the latest time that the individual with the larger chute could safely pull the ripcord?

4. **It's sabotage.** Two of the parachutes for the average customer have been tampered with. They have small holes in them which will affect their ability to provide wind resistance. For parachute A, $k = 12e^{-0.001t} + 0.75$, and for parachute B, $k = 12e^{-0.005t} + 0.75$. Solve the system of equations for both parachutes. Will the hole in either parachute affect the safety of the jump (that is, will the velocity when the jumper reaches the ground be greater than the desired 20 feet per second)?

Project^{TE}: It's Just a Disease

The Organization Who Investigates Epidemics (OWIE) has discovered a dangerous new disease called Elephant Pox and has hired you to help create a model to help predict the spread of the disease. A vaccine is being developed for Elephant Pox, so it is recommended that you use an SVIR model for the predictions. SVIR models separate the overall population into four classes: Susceptible (S), Vaccinated (V), Infected (I), and Recovered (R). Susceptible people become infected at a rate (all rates are in terms of people per day) of $\frac{\alpha SI}{N}$, where α is the infection rate parameter, and N is the total population. Vaccinated people become infected at a rate of $\frac{\alpha(1-\sigma)SI}{N}$, where σ represents the efficacy of the vaccine. Infected people recover at a rate of γI, where γ is the recovery rate.

1. **If you build it, you can predict.**

   (a) Draw a schematic to represent the dynamics of the SVIR model.

   (b) Using your schematic, create a system of differential equations that can be used to model the spread of the disease.

2. **But I'm afraid of needles.** OWIE has reported that Elephant Pox has a 95% infection rate (α), its vaccine is 98% effective (σ), and infected individuals recover in about 15 days ($\gamma = 1/15$). One of the reasons that preventable diseases persist is because a large portion of the people who are able to be vaccinated refuse to do so. Use your model to predict the spread of Elephant Pox in a population of 100,000, if only 50% of the population

is initially vaccinated and there is initially 1 infected individual. Plot your
results. What happens to the susceptible population?

3. **None of us is as strong as all of us.** Herd immunity is a concept that
   prescribes that if a high enough percentage of the population gets vaccinated
   against a disease, then those individuals who are not able to be vaccinated
   will be protected from the disease based on the vaccinated population. Use
   the model to predict the spread of Elephant Pox in a population of 100,000 if
   95% of the population is initially vaccinated, and there is initially 1 infected
   individual. Plot your results. What happens to the susceptible population.

4. **That's a disease of a different color.** OWIE has identified another disease, Horse Fever, which can also be vaccinated against. Horse Fever can be
   modeled with the same SVIR model, but with different parameter values.
   OWIE has reported that Horse Fever has an 85% infection rate (α), its vaccine is 92% effective (σ), and infected individuals recover in about 12 days
   ($\gamma = 1/12$). Consider a population of 100,000 with 1 infected individual.
   Use your model with different initial conditions (change the percent of the
   individuals who were initially vaccinated) to get an estimate for the level of
   vaccination that would provide herd immunity against Horse Fever.

4.4   Second Order Differential Equations

There are many physical and biological processes whose dynamics cannot be adequately modeled with a simple first order differential equation. For those processes, a second (or higher) order differential equation can be derived to represent the changing system. Often, the derivation of these models relies on well known physical properties or laws which are referred to as *first principles*. In this section, we will turn our attention to the derivation of a second order model, and discuss analytical and numerical solutions for such models.

Constructing a Second Order Differential Equation Model

In deriving differential equation models of higher order, we will consider a state variable of interest and formulate an equation based on the state variable and its derivatives which dictate the dynamics of the desired system. The derivation of a higher order model is more complicated than those of first order models which we have already seen. As an example of how to construct a higher order model, we will derive the differential equation of motion for a spring-mass system.

Consider a mass that is suspended vertically in air by a flexible spring (see Figure 4.13 for details). We'd like to derive a model for the position of the mass, x, with respect to time, t, if the system were put into motion. Note that the schematic in Figure 4.13 depicts the mass at its equilibrium point, x_0. At the equilibrium point, the forces acting upon the mass are equal, resulting in no movement in the system. For this derivation, we will assume that positions below the equilibrium point will have positive value, and positions above the equilibrium point will have negative values. Under this assumption, we can define the total force on the mass to be equal to the force due to gravity minus the force due to the spring. When the mass is at the equilibrium point (x_0),

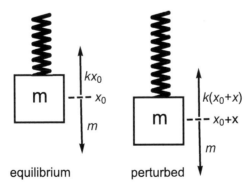

FIGURE 4.13: Schematic of a mass suspended vertically by a flexible spring.

the total force exerted on the mass is equal to zero. The force due to gravity is expressed by using *Newton's Second Law of Motion*. Newton's Second Law states that the force on a moving object is equal to the mass of the object multiplied by the acceleration of the object. So, the force due to gravity can be expressed as mg, where m is the mass of the suspended object, and g is the acceleration due to gravity.

Similarly, we can derive an expression for the force due to the spring. *Hooke's Law* states that the force exerted by a spring is equal to the spring constant for the spring multiplied by the displacement of the object. For our system, the force due to the spring can be represented as kx_0, where k is the spring constant, and x_0 represents the displacement at equilibrium. Putting these together, we'll get the following equation for the total force on the system at equilibrium:

$$0 = mg - kx_0. \tag{4.14}$$

Now, let's consider what happens when the system is perturbed or set into motion. Assume that the mass is further displaced by a value of x. Then the total displacement would be $x_0 + x$. The additional displacement will move the mass from its equilibrium point, resulting in a nonzero force on the system. Again, this force can be computed by considering the force due to gravity minus the force imparted by the spring. For the perturbed system, we'll have the following equation:

$$F = mg - k(x_0 + x).$$

Using Newton's Second Law of Motion, we can rewrite F as ma, yielding

$$ma = mg - k(x_0 + x).$$

From Calculus I, we know that acceleration can be represented as the second derivative of displacement with respect to time. Applying this principle and distributing k on the right hand side leads us to

$$m\frac{d^2x}{dt^2} = mg - kx_0 - kx.$$

Now, we know from (4.14) that the first two terms on the right hand side will equal zero. Thus, our equation simplifies to

$$m\frac{d^2x}{dt^2} + kx = 0. \tag{4.15}$$

Equation (4.15) is known as the differential equation for free undamped motion. The motion is classified as free because there is no external force working upon the system. The motion of the suspended mass is based solely on the force due to gravity and the force due to the spring.

Using a similar process, we could also derive the differential equation for damped motion by assuming that there is a damping force, $\beta\frac{dx}{dt}$, that acts opposite to the direction of motion, yielding

$$m\frac{d^2x}{dt^2} + \beta\frac{dx}{dt} + kx = 0. \tag{4.16}$$

The differential equations for forced motion would be analogous to (4.15) and (4.16), except the right hand side of the equations would be replaced with some forcing function that drives the motion of the corresponding systems.

Solving a Second Order Differential Equation

Once a suitable model has been derived or chosen, it is desirable to find the solution which can be used to represent the modeled system. Again, we will discuss finding the solution for a second order differential equation analytically, as well as numerically. We'll first consider analytical solutions for second order differential equations.

Example 4.4.1 *Dr. Roger VonSteen, the physics teacher at Hypothetical High School, is showing his class a demonstration of harmonic motion. He attaches a 6 kg mass to a spring which stretches $\frac{49}{65}$ meters to reach its equilibrium point. The mass is then connected to a device that will impart a downward force of $12\cos t$ Newtons to the mass. If the air resistance is equal to 36 times the instantaneous velocity, and the mass is started from the equilibrium position with a downward velocity of 2 m/s, find the equation of motion for Dr. VonSteens example. Create a graph that illustrates the motion for the first 20 seconds. Assume that the acceleration due to gravity is 9.8 m/s^2.*

Since we are dealing with a forced, damped spring-mass system, we will need to set up the following differential equation:

$$m\frac{d^2x}{dt^2} + \beta\frac{dx}{dt} + kx = F(t). \tag{4.17}$$

From the information given about Dr. VonSteen's demonstration, we know that the mass (m) is 6 kg, and the damping coefficient (β) will be 36 kg/s. Also, the forcing function $(F(t))$ is $12\cos t$ and we have initial conditions $x(0)=0$ m and $x'(0) = 2$ m/s. We were given all of the necessary information except for the spring constant k. We can find k by using Hooke's Law along with the information about the initial stretching of the spring. Remember, Hooke's Law states that the force imparted on a spring is equal to the spring constant (k) times the displacement of the spring. In our demonstration, the force imparted on the spring is the weight of the attached mass $(F = mg)$, so we can set up the following equation to find a value for the spring constant k:

$$\begin{aligned} mg &= kd \\ \Rightarrow 6(9.8) &= \frac{49}{65}k \\ \Rightarrow k &= 78. \end{aligned}$$

Now that we have found the value for the spring constant, we can set up the initial value problem by substituting our known quantities into (4.17), yielding the following:

$$6\frac{d^2x}{dt^2} + 36\frac{dx}{dt} + 78y = 12\cos t; \qquad x(0) = 0, \frac{dx}{dt}(0) = -2.$$

Note that we can simplify the equation by dividing each term by 6, yielding

$$\frac{d^2x}{dt^2} + 6\frac{dx}{dt} + 13y = 2\cos t; \qquad x(0) = 0, \frac{dx}{dt}(0) = -2.$$

We'll first find the analytical solution by hand using methods covered in an introductory differential equations class. We start by finding the complementary function (x_c), or solution to the homogeneous part of the differential equation. To find x_c, we'll examine the auxiliary equation $m^2 + 6m + 13 = 0$. The roots of the auxiliary equation are $m = -3 \pm 2i$, which leads to a complementary function of

$$x_c = e^{-3t}\left(c_1 \cos 2t + c_2 \cos 2t\right).$$

We'll now find the particular solution (x_p) to the differential equation. Since the right hand side of our equation is $2\cos t$, we'll assume that x_p takes the same general form. That is, $x_p = A\cos t + B\sin t$. None of the terms of our proposed particular solution repeat the terms in x_c, so we can proceed with the proposed x_p. To find the values for A and B, we substitute x_p and its derivatives into the differential equation. First, we'll find the following derivatives:

$$\begin{aligned} x_p &= A\cos t + B\sin t \\ \frac{dx_p}{dt} &= -A\sin t + B\cos t \\ \frac{d^2x_p}{dt^2} &= -A\cos t - B\cos t. \end{aligned}$$

Substituting these functions into the differential equation yields the following:

$$\begin{aligned} \frac{d^2x}{dt^2} + 6\frac{dx}{dt} + 13y &= 2\cos t \\ \Rightarrow -A\cos t - B\cos t + 6(-A\cos t - B\cos t) + 13(A\cos t + B\sin t) &= 2\cos t \\ \Rightarrow (12A + 6B)\cos t + (12B - 6A)\sin t &= 2\cos t. \end{aligned}$$

We can solve for A and B by equating the coefficients of each side of the equation. That is, the coefficient of $\cos t$ on the right hand side must equal the coefficient of the $\cos t$ term on the left hand side, and the same for the $\sin t$ terms. This leads us to the following system of equations:

$$\begin{aligned} 12A + 6B &= 2 \qquad (\cos t \text{ coefficients}) \\ -6A + 12B &= 0 \qquad (\sin t \text{ coefficients}). \end{aligned}$$

Solving the system of equations leads us to $A = \frac{2}{15}$ and $B = \frac{1}{15}$. Plugging these into x_p, and adding it to x_c will give us the following solution to the differential equation:

$$x(t) = e^{-3t} \left(c_1 \cos 2t + c_2 \sin 2t \right) + \frac{2}{15} \cos t + \frac{1}{15} \sin t.$$

The final step to find the solution to our initial value problem is to apply the initial conditions to find values for the arbitrary constants c_1 and c_2. We start with the initial condition for the displacement, $x(0) = 0$, yielding

$$
\begin{aligned}
0 &= e^{-3(0)} \left(c_1 \cos 2(0) + c_2 \sin 2(0) \right) + \frac{2}{15} \cos 0 + \frac{1}{15} \sin 0 \\
\Rightarrow 0 &= c_1 + \frac{2}{15} \\
\Rightarrow c_1 &= -\frac{2}{15}.
\end{aligned}
$$

Applying the initial condition for the velocity, $\frac{dx}{dt}(0) = -2$, we can find the following value for c_2:

$$
\begin{aligned}
x'(t) &= -3e^{-3t} \left(c_1 \cos 2t + c_2 \sin 2t \right) + 2e^{-3t} \left(-c_1 \sin 2t + c_2 \cos 2t \right) \\
&\quad - \frac{2}{15} \sin t + \frac{1}{15} \cos t \\
\Rightarrow x'(0) &= -3e^{-3(0)} \left(c_1 \cos 2(0) + c_2 \sin 2(0) \right) + 2e^{-3(0)} \left(-c_1 \sin 2(0) \right. \\
&\quad \left. + c_2 \cos 2(0) \right) - \frac{2}{15} \sin 0 + \frac{1}{15} \cos 0 \\
\Rightarrow -2 &= -3c_1 - 2c_2 + \frac{1}{15} \\
\Rightarrow -2 &= -3 \left(-\frac{2}{15} \right) - 2c_2 + \frac{1}{15} \\
\Rightarrow c_2 &= -\frac{37}{30}.
\end{aligned}
$$

This will lead us to the following equation of motion for Dr. VonSteen's demonstration:

$$x(t) = e^{-3t} \left(-\frac{2}{15} \cos 2t - \frac{37}{30} \sin 2t \right) + \frac{2}{15} \cos t + \frac{1}{15} \sin t.$$

We could also find the analytical solution using *Mathematica*. The following command will allow us to solve the initial value problem in *Mathematica*.

In1 := sol = DSolve[{x''[t] + 6 * x'[t] + 13 * x[t] == 2 * Cos[t],x[0] == 0,
x'[0] == -2},x[t],t]
FullSimplify[%]

The previous command will generate the following output.

$$\text{Out1} \quad := \quad \left\{ x[t] \to \frac{1}{60}(-8\text{Cos}[2t] + 3e^{3t}\text{Cos}[t]\text{Cos}[2t] + 5e^{3t}\text{Cos}[2t]\text{Cos}[3t] \right.$$

$$-9e^{3t}\text{Cos}[2t]\text{Sin}[t] - 74\text{Sin}[2t] + 9e^{3t}\text{Cos}[t]\text{Sin}[2t] + 5e^{3t}\text{Cos}[3t]$$

$$\text{Sin}[2t] + 3e^{3t}\text{Sin}[t]\text{Sin}[2t] - 5e^{3t}\text{Cos}[2t]\text{Sin}[3t] + 5e^{3t}\text{Sin}[2t]$$

$$\left. \text{Sin}[3t]) \right\}$$

$$\text{Out2} \quad := \quad \left\{ x[t] \to \frac{1}{15} \left(2\text{Cos}[t] + \text{Sin}[t] - e^{-3t} \left(2\text{Cos}[2t] + 37\text{Cos}[t]\text{Sin}[t] \right) \right) \right\}$$

Note that if we rewrite the final term from the *Mathematica* output using the double angle formula, then it will exactly match the solution that we found above. Now, we will graph the solution to the differential equation for 20 seconds. We'll use the following commands in *Mathematica*.

$$\text{In3} \quad := \quad \text{Plot}[x[t]/.\text{sol},7\{t,0,20\},\text{PlotStyle} \to \text{Black},$$
$$\text{AxesLabel} \to \{ \text{``time (sec)''}, \text{``displacement (m)''} \}]$$

Figure 4.14 represents the motion of the spring demonstration from Dr. VonSteen's class.

■

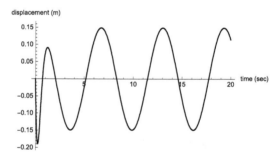

FIGURE 4.14: Displacement of the spring-mass system set up for Dr. Von-Steen's demonstration over 20 seconds.

Now, we'll discuss numerical solutions for second order differential equation models. Consider the following example.

Example 4.4.2 *One evening, while helping his parents clean their attic, Jon ran across one of his old toys. Charley in the Box was his favorite toy when he was 4 years old. After turning a crank, a plastic head attached to a spring would pop out of the box and bounce up and down. While playing with the old toy, Jon realized that it didn't work in the same manner as it did so many years* *ago. When the toy was new, the movement of Charley's head could be modeled by the differential equation*

$$\frac{d^2x}{dt^2} + \frac{dx}{dt} + 10x; \qquad x(0) = -5, \frac{dx}{dt}(0) = 1. \tag{4.18}$$

After years of sitting in the attic, the spring within the toy has deteriorated, and its spring constant decays over time. The movement of Charley's head is now represented by

$$\frac{d^2x_1}{dt^2} + \frac{dx_1}{dt} + 10e^{-0.5t}x_1; \qquad x_1(0) = -5, \frac{dx_1}{dt}(0) = 1. \tag{4.19}$$

Jon would like to compare the motion of the toy to its previous state. Numerically solve the above equations and plot them on the same axis for 10 seconds to facilitate the comparison of the toy's current state with its previous performance.

Again, we'll compute the numerical solutions using both *Mathematica* and *MATLAB*. Starting with *Mathematica*, we'll use the following commands to solve and plot both solutions together.

In1 := sol1=NDSolve[$\{x''[t] + x'[t] + 10 * x[t] == 0, x[0] == -5, x'[0] == 1\}$, $x[t], \{t,0,10\}$]

yields the following output.

Out1 := $\{\{x[t] \rightarrow \text{InterpolatingFunction}[...][t]\}\}$

The second equation is solved with the command

In2 := sol2=NDSolve[$\{x1''[t] + x1'[t] + 10 * Exp[-.5t] * x1[t] == 0$, $x1[0] == -5, x1'[0] == 1\}, x1[t], \{t,0,10\}$]

yielding the following output.

Out2 := $\{\{x1[t] \rightarrow \text{InterpolatingFunction}[...][t]\}\}$

We can plot both solutions using the following command.

In3 := Plot[Evaluate[$\{x[t]/.sol1, x1[t]/.sol2\}], \{t,0,10\}$,PlotStyle $\rightarrow$ {Black, Red},AxesLabel $\rightarrow$ {"time (sec)", "position (in)"}]

The resulting graph is displayed in Figure 4.15(a).

In order to solve a second order differential equation in *MATLAB*, we'll have to do some extra preparation. *MATLAB*'s differential equation solvers are designed to solve first order differential equations. So, we'll need to transform our second order differential equation into a system of first order differential equations. To accomplish this, we'll introduce a "dummy" variable. Starting with the equation for the toy when it was new, let $w = \frac{dx}{dt}$. If we substitute this into (4.18), we'll have the following.

$$\frac{dw}{dt} + w + 10x = 0 \Rightarrow \frac{dw}{dt} = -w - 10x.$$

This results in the following system of first order equations.

$$\frac{dx}{dt} = w \tag{4.20}$$

$$\frac{dw}{dt} = -w - 10x. \tag{4.21}$$

We can use the same approach to rewrite (4.19) as the following system of first order differential equations.

$$\frac{dx_1}{dt} = w_1 \tag{4.22}$$

$$\frac{dw_1}{dt} = -w_1 - 10e^{-0.5t}x_1. \tag{4.23}$$

Now that we have rewritten both equations as systems, we can create a *MAT-LAB* file to solve each system and plot their solutions.

```
****************************************
%Initialize time vector and Initial Condition(s)
tspan=0:.01:10;  %Define length of time (0 to 10 seconds)
%Initial Conditions
x0=-5;    %initial position
v0=1;     %initial velocity
IC=[x0;v0];
%Call ODE solver
[t,y]=ode45(@OldCharley,tspan,IC);
x=y(:,1); w=y(:,2);
[t1,y1]=ode45(@PresentCharley,tspan,IC);
x1=y1(:,1); w1=y1(:,2);
%plot results
figure
plot(t,x,'k',t1,x1,'r');
xlabel('time (seconds)');
ylabel('position (inches)');
legend('Past','Present')
%-------------------------------------------------------------
function dy=OldCharley(~,y)
%Define States
x=y(1); w=y(2);
%Define ODES
```

```
dy=[w; -w-10*x];
end
%------------------------------------------------
function dy=PresentCharley(t,y)
%Define States
x=y(1); w=y(2);
%Define ODES
dy=[w; -w-10*exp(-0.5*t)*x];
end
**********************************
```

The *MATLAB* code produces the results that are depicted in Figure 4.15(b). ∎

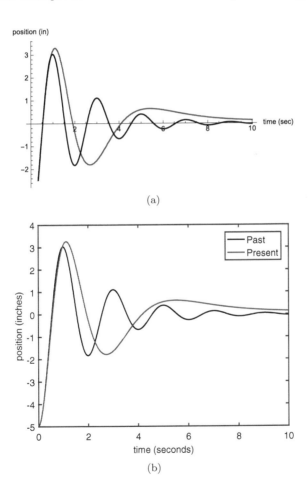

(a)

(b)

FIGURE 4.15: Simulations of the motion of Charley in the Box from the past (black) and present (red) as computed by (a)*Mathematica* and (b) *MATLAB*.

Exercises

1. The Wamko toy company has developed a new toy which they have called the Self-Propelled Yo-Yo. They claim that once started, the toy will continue to oscillate until it is caught. The path of motion of the toy can be modeled using an undamped free mass-spring system.

   (a) Assume that the mass of the toy is .5 kg, and that the spring constant is 8 N/m. If the toy is started at a position of .5 m above its equilibrium with a downward velocity of 1 m/s, find an analytical solution for the equation of motion for the toy.

   (b) ^{TE} Plot the position of the toy over 120 seconds. Does it appear to be slowing at all?

2. George (being very curious) wanted to see if the Self-Propelled Yo-Yo works underwater. He takes the toy described in Exercise 1 into his family's swimming pool. The water in the pool imparts a damping force of 2 times the instantaneous velocity. He again starts the toy at a position of .5 m above its equilibrium with a downward velocity of 1 m/s.

   (a) Find an analytical solution for the equation of motion for the toy as it moves underwater.

   (b) ^{TE} Plot the position of the toy over 120 seconds. Does the toy oscillate underwater? If so, when do the oscillations subside?

3. ^{TE} The differential equation $mx'' + kx + k_1x^3 = F_0 \cos \omega t$ is the equation for a Duffing oscillator. Consider the initial value problem $x'' + 2x + k_1x^3 = 7 \cos t, x(0) = 2, x'(0) = 0$.

   (a) Find a numerical solution for the initial value problem for $k_1 = 0.1, 1, 10$, and 100, with t from 0 to 5. Plot the solution curves on the same axes.

   (b) What happens to the solution of the Duffing oscillator as k_1 increases?

4. ^{TE} Bipolar II disorder is characterized by a combination of hypomania and depressive episodes. Daugherty et. al [25] presented a mathematical model based on the van der Pol oscillator that could represent the change in mood of someone suffering from bipolar disorder. Consider the second order differential equation

$$\frac{d^2x}{dt^2} - \alpha\frac{dx}{dt} - \omega^2 x - \beta x^2\frac{dx}{dt} = \gamma x^4\frac{dx}{dt},$$

where x represents the emotional state of the patient, $\frac{dx}{dt}$ is the rate of mood changes between hypomania and severe depression, and $\alpha > 0$ and ω are parameters. The β and γ terms in the model represent treatment.

(a) Find and plot a numerical solution to the differential equation for 50 years, when $\alpha = 0.36, \beta = -100, \omega = 5$, and $\gamma = 0$. Use initial conditions $x(0) = 0.12, \frac{dx}{dt}(0) = 0$. This parameterization represents an untreated individual. Explain what the model suggests is happening to the patient's mood over time.

(b) Find and plot a numerical solution to the differential equation for 50 years, when $\alpha = 0.1, \beta = -100, \omega = 5$, and $\gamma = 5000$. Use initial conditions $x(0) = 0.18, \frac{dx}{dt}(0) = 0$. Explain what the model suggests is happening to the patient's mood as time increases. How does this compare to the untreated case?

5. With further investigation, Daugherty et. al [25] determined that the equation in Exercise 4 had several drawbacks to it. They proposed a second model using the following differential equation.

$$\frac{d^2x}{dt^2} + f(x)\frac{dx}{dt} + h(x) = \rho\left(\frac{dx}{dt}\right)^3 + \mu\left(\frac{dx}{dt}\right)^5 + \nu\left(\frac{dx}{dt}\right)^{11},$$

where the treatment is represented by the function on the right hand side of the equation.

(a) Let $f(x) = -0.38, h(x) = 180x, \mu = 0.78, \nu = -0.00093$, and $\rho = 0.38$. This parameterization describes an individual with steadily worsening mood swings throughout his childhood and adolescent years. Find and plot a numerical solution using this parameterization for time equal to zero to 20. Use initial conditions $x(0) = 0.001, \frac{dx}{dt} = 0$.

(b) Changing the parameter ρ drastically affects the oscillations of the model. To simulate an individual on treatment, let $\rho = -3.302$. Using the previous parameterization, we saw dramatic mood swings develop around age 20. Assume that once these oscillations were noticed, treatment was started. Find and plot a numerical solution to the model using the parameters defined above with $\rho = -3.302$ for time equal to 20.35 to 65. Use initial conditions $x(20.35) = -0.1, \frac{dx}{dt} = -2.5$. Explain what the results indicate about the mood oscillations of the treated individual.

Project^{TE}: Through the Air with the Greatest of Ease

The What Was I Thinking© Extreme Sports Company wants to expand their business to offer bungee jumping and a Skycoaster. They have again asked for your assistance in planning the details of these activities.

1. **What goes down comes back up, and then down again...** For the bungee jumping attraction, customers will jump off a platform that is 20 feet in the air with an initial velocity of 0 feet per second. Customers will wear a special suit that will impart air resistance that is equal to 5 times the instantaneous velocity. The bungee cord has a spring constant of 120 pounds per foot. You have determined that the motion of a person on the bungee cord can be modeled with the following differential equation.

$$m\frac{d^2y}{dt^2} + 5\frac{dy}{dt} + 120y = 1250,$$

where $y(t)$ is the height of the jumper above the ground at time t, and m is the mass of the jumper.

(a) The company wants to be sure that the bungee jump will be safe for customers of mass up to 200 kg. Solve the differential equation and plot the results for $m = 85, 125, 165$, and 205. Does the bungee event appear to be safe for customers of varying masses? What appears to be the difference in the motion of the bungee jump for the different customers?

(b) The manufacturer of the bungee cord no longer makes bungee cords with the specifications of the one that the company will initially use. Eventually, the bungee cord will need to be replaced. They offer cords with effective spring constants of $k = 85, 110, 135$ or 150. Consider a customer with a mass of 125 kg. Which of the bungee cord options will allow the customer to get closest to the ground without hitting it? How high from the ground would the customer reach on the initial fall?

2. **Swing set, swing set, back and forth we go.** For the Skycoaster attraction, customers will be attached to an 18 ft cord that is attached to a 22 ft tower. Customers have been hoisted into the air, where the cord makes an angle of $\pi/3$ with the tower. When the customer is dropped from the raised position, they have an initial angular velocity of $-1/2$ radians per second. The customer will then swing back and forth until they come to a stop. You have determined that the motion of the Skycoaster can be modeled with the following differential equation, which deals with the motion of a pendulum.

$$\frac{d^2\theta}{dt^2} + \beta\frac{d\theta}{dt} + \frac{g}{L}\theta = 0,$$

where $\theta(t)$ represents the angle that the cord makes with the vertical tower ($\theta = 0$ will be where the customer stops, even with the tower), β is the damping or air resistance coefficient, g is the acceleration due to gravity, and L is the length of the cord.

(a) Solve the differential equation and plot the equation of motion.

(b) The company is considering offering different lengths of cords for the Skycoaster. Solve the differential equation for $L = 10, 12, 15$, and 20. What affect does the length of the cord have on the duration of the "ride"?

3. **I've got to get my money's worth.** Ben only has enough money to ride one of the new attractions. He has a mass of 105 kg. He wants to get the most out of his money, so he would like for you to determine which of the rides would last longer. Use both models to determine how long it takes for the two different "rides" to stop oscillating. For which ride should Ben purchase a ticket?

Project^{TE}: Series Circuit Analysis

The purpose of this project is to develop and analyze a mathematical model of an LRC circuit. (See the diagram below.) The circuit consists of a voltage source, an inductor of inductance L, a resistor of resistance R, and a capacitor of capacitance C. At some instant, which we take to be time 0, the switch is closed, completing the circuit. Assume the charge on the capacitor is initially 0, and the current is initially 0.

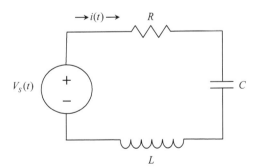

The voltage source, V_S, is considered to be the input to the system, and the voltage measured across the capacitor, V_C, is considered to be the output. You will study the repsonse of this system to a constant applied voltage and to a sinusoidal applied voltage. For the project description and report, the following notation should be used.

t	time (seconds)
q	charge on the capacitor at time t (Coulombs)
i	current in the circuit at time t (Amperes)
V_S	Source of the voltage applied to the circuit (Volts)
V_C	voltage across the capacitor (Volts)
L	inductance of the inductor (henrys)
C	capacitance of the capacitor (farads)
R	resistance of the resistor (Ohms, Ω)

Problem Statement:

The project consists of the following steps.

1. **Formulation of the model.** Use the following information to formulate an initial value problem modeling this circuit. Take t to be the independent variable, and q to be the dependent variable.

   - Kirchoff's Second Law states that the sum of the voltage across the inductor, the resistor, and the capacitor equals the applied voltage V_S.

   - The voltage drop across the inductor equals $L\dfrac{di}{dt}$

   - The voltage drop across the resistor equals $R \cdot i$.

   - The voltage drop across the capacitor equals q/C.

   - $i = \frac{dq}{dt}$.

   - The charge on the capacitor is initially 0 Coulombs.

   - The initial current is equal to 0 Amperes.

2. **Response to a constant applied input.**

   (a) Solve the initial value problem constructed in part 1 assuming that the circuit has inductance of 10 henrys, resistance of 4 Ω, capacitance of 0.2 farads, and input voltage $V_S = 20$ V.

   (b) Show that q approaches a constant value as $t \to \infty$.

   (c) Examine the effect that the inductance has on the solution, q. How does the solution to the initial value problem change if L is gradually increased? How does the solution change if L is gradually decreased?

   (d) Examine the effect that the resistance has on the solution, q. How does the solution to the initial value problem change if R is gradually increased? How does the solution change if R is gradually decreased?

3. **Response to sinusoidal input.**

   (a) Solve the initial value problem from part 1 assuming that the circuit has inductance of 10 henrys, resistance of 4Ω, capacitance of 0.2 Farad, and a sinusoindal input voltage $V_S = 20\cos(\omega t)$ V. You may want to use a computer algebra system or other technology to assist with the integration in the solution to the problem.

   (b) Show that the response q from part 3a contains a transient term q_{tr} that approaches zero as $t \to \infty$ and a steady-state term q_{ss} that does not approach zero.

   (c) Create and compare the following graphs. How does the steady state charge change as ω increases? What do you think will happen as $\omega \to \infty$? This LRC circuit acts as a low-pass filter, meaning that it filters out high-frequency signals. Do your results support this statement?

      i. Take $\omega = 0.1$. Plot the charge q_S associated with the sinusoindal input V_S (remember, $C = \frac{q}{V}$) and the steady state response q_{ss} on the same set of axes for $0 \le t \le 60\pi$.

      ii. Take $\omega = 0.3$. Plot the charge q_S associated with the sinusoindal input V_S (remember, $C = \frac{q}{V}$) and the steady state response q_{ss} on the same set of axes for $0 \le t \le 20\pi$.

      iii. Take $\omega = 0.5$. Plot the charge q_S associated with the sinusoindal input V_S (remember, $C = \frac{q}{V}$) and the steady state response q_{ss} on the same set of axes for $0 \le t \le 12\pi$.

      iv. Take $\omega = 1$. Plot the charge q_S associated with the sinusoindal input V_S (remember, $C = \frac{q}{V}$) and the steady state response q_{ss} on the same set of axes for $0 \le t \le 6\pi$.

      v. Take $\omega = 2$. Plot the charge q_S associated with the sinusoindal input V_S (remember, $C = \frac{q}{V}$) and the steady state response q_{ss} on the same set of axes for $0 \le t \le 3\pi$.

      vi. Take $\omega = 5$. Plot the charge q_S associated with the sinusoindal input V_S (remember, $C = \frac{q}{V}$) and the steady state response q_{ss} on the same set of axes for $0 \le t \le 1.2\pi$.

Chapter Synthesis: Independent Investigation

1. ^{TE} (Revisit Problem 5 from Section 4.2) You previously considered the dynamics of a population of trout in a local pond as depicted in the schematic in Figure 4.5, but fishing will be more popular in certain months than others. Recreate the differential equation representing the figure (as in the previous problem), and assume that the birth rate constant (α) is 2 per month and the death rate constant is .75 per trout per month. Then formulate a model for the restocking and harvesting rates based on the time of year. Assume $t = 0$ represents the beginning of the year, but assume your initial condition is that 2000 fish are in the pond (or that T=2) at the beginning of a month that fishers will be heavily active (in other words, not January). Your goals include:

   - Keeping the population of trout from dying out.
   - Keeping the population of trout below 2500.
   - Preserving the ability of fishers to catch trout (for tourism). You want to be able to have significant harvesting in some way.

   Make recommendations that indicate the rates at which fish should be restocked and what times of year the pond should be open for fishing.

2. ^{TE} Section 4.2 contains a project on "Crime Scene Investigation: Differential Equations Unit." Consider a similar situation where you are planning the murder. Create two different scenarios that produce the same body temperature at a particular time the body is found. Assume that you will tip off the authorities with a well-timed call.

3. ^{TE} (Revisiting Example 4.3.2 from Section 4.3) In a previous example, a system of two differential equations was formed to describe Happy Cow Dairy's mixing of milk and chocolate in two reservoirs, one upper and one lower. In the previous example, the system begins with chocolate in the upper reservoir and no chocolate in the lower reservoir. However, after the mixing has begun at the beginning of the workday, Happy Cow Dairy would like to harvest the perfect chocolate milk mixture (which they believe is somewhere between 0.7% and 1% chocolate) from the lower reservoir to sell but continue to mix more chocolate milk. Happy Cow Dairy can remove up to 1000 gallons from the lower reservoir (and immediately replace that chocolate milk with plain milk) at the same time they add some amount of chocolate in pounds to the upper reservoir. Develop a plan for how Happy Cow Dairy can get as much chocolate milk during the 8 hour workday as possible, including when and how much milk they should harvest, how often

they should remove chocolate milk, how much chocolate they should start with, and how much new chocolate they should add with each removal.

4. ^{TE} (Revisiting Example 4.3.1 from Section 4.3) In the gang population example, a system of three equations was created to represent the population dynamics of the general population of a town with the populations of two rival gangs. Consider the following:

   (a) Solve the system of differential equations with a variety of parameter values. Discuss what happens to the three populations as you increase or decrease each parameter value. You may start with the following parameter values for your analysis: $\alpha_D = 0.2, \alpha_P = 0.1, \omega_D = 0.05, \omega_P = 0.075, \mu_D = 0.2, \mu_P = 0.1, \delta_D = 4 \times 10^{-4}, \delta_P = 7 \times 10^{-4}, G(0) = 50{,}000, D(0) = 1000$, and $P(0) = 500$.

   (b) Modify the model to include a popluation of police officers who are recruited from the general population. Members of each gang are removed from the model (sent to jail) at a rate that is proportional to the product of the gang population and the population of police officers. Find a parameterization for this new model which will lead to the removal of the gangs from the community.

5. (Revisiting Example 1.1.1 From Section 1.1) In Section 1.1, there was an example that dealt with a radioactive material that was in decay. You were told that the initial amount of the chemical is 100 g, and the half-life of the chemical is 3 years. Formulate a first order differential equation to represent the decay of the chemical. Show that the solution of your differential equation is the same as the equation presented in the example.

6. ^{TE} (Revisiting Exercise 4 from Section 2.5) In Exercise 4, a system of sequences was presented to represent the populations of a species of predator and prey. Consider the following:

   (a) Use the given sequences to generate the population of the predators and prey for 200 steps.

   (b) Formulate a corresponding system of differential equations that could be used to model the interactions of this predator-prey system.

   (c) Solve the system of differential equations for a time period that would be equivalent to the 200 steps from the sequence.

(d) Compare the solutions from the sequence model to the solutions from the differential equations model. Are there any differences between the solutions?

5

Additional Topics in Modeling

Goals and Objectives

The following chapter is written toward students who have some computing background. There is no assumption related to the programming language being used by the student and many of the problems are posed in terms of a flowchart or pseudocode. It is important to note that there are exercises and projects, in this chapter, that require programming software or a computer algebra system and the ability to generate random numbers. Technology heavy exercises and projects are marked with a ^{TE} at the beginning and are not software dependent.

Goals:

- Section 5.1 (Monte Carlo Simulation)

    1. To introduce the student to the concept of the Monte - Carlo simulation method and randomness in modeling.

    2. To expose the student to the importance of multiple simulation runs, similar to multiple samples, to explain typical or average behavior.

    3. To expose the student to the practice of developing flowcharts and pseudocode that explain the steps of a simulation.

    4. To introduce the student to a variety of applications that can benefit from simulation modeling.

- Section 5.2 (Equation-Based Models versus Agent-Based Models)

    1. To introduce the student to the basic ideas of difference equation modeling.

    2. To introduce the student to the concept of agent-based modeling.

    3. To discuss studies related to comparing equation-based modeling and agent-based modeling.

- Section 5.3 (Modeling Using Game Theory): To introduce the basic concepts and applications of game theory, including payoff matrices and pure and mixed strategy equilibria.

- Section 5.4 (Voronoi Diagrams)

    1. To introduce the concepts of Delaunay Triangulation and Voronoi Diagrams.

    2. To expose students to optimization problems in which Voronoi Diagrams can be applicable.

5.1 Monte Carlo Simulation

Sometimes, in real life situations, it is difficult to perform an experiment to collect data and thus you must run a simulation to model the situation to obtain results. *Monte Carlo methods* (or *Monte Carlo* experiments) are a broad class of computational algorithms that rely on repeated random sampling to obtain numerical results. In this chapter, we will see a few situations of how Monte Carlo simulations can be valuable. We begin with an interesting experiment introduced by Georges Louis LeClerc, Count of Buffon, in 1777, called Buffon's Needle [35].

Example 5.1.1 *(Buffon's Needle) In this example, we present a simplified version of the Buffon's Needle problem. Given a toothpick of length 2.5 inches and a plane with parallel lines that are 1 inch apart, what is the probability that the toothpick will cross one of the lines?*

Take 100 toothpicks, any large number will work, and drop them on your plane. Count the number of toothpicks crossing a line. Repeat this process many times and take the average for an approximation of the probability. So what is the exact probability?

Suppose that the toothpick lands at an angle of θ, which we will assume is between 0 and π, based on symmetry. Now draw a line parallel to lines on the plane that intersects the center of the toothpick. Let d be the distance from the center of the toothpick to the closest line on the plane. If $x \geq d$, seen in Figure 5.1, then the toothpick crosses the parallel line; otherwise the toothpick does not cross the line.

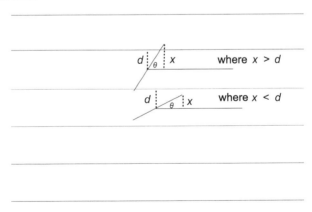

FIGURE 5.1: Example of Buffon Needle simulation.

If $x \geq d$ then from trigonometry we can see that $0 \leq d \leq 1.25 \sin(\theta)$. Thus, we need to find out the probability of this happening. Recall that in general the probability of an event occurring is the number of ways the event can occur divided by the total number of possible outcomes. In this case, θ can take on any value between 0 and π and thus the total possible outcomes for d is 1.25π. Thus, in order to find this probability, we wish to find the area under the curve $1.25\sin(\theta)$ for $0 \leq \theta \leq \pi$ with respect to the area of the rectangle 1.25π.

$$P(0 \leq d \leq 1.25\sin(\theta)) = \frac{\int_0^\pi 1.25\sin(\theta)d\theta}{1.25\pi} = \frac{2}{\pi}.$$

In addition, we need $d < \frac{1}{2}$ since the lines are 1 inch apart and if d is

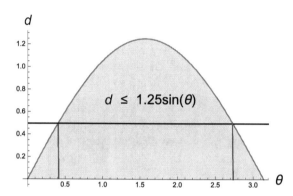

FIGURE 5.2: Area representing probability in Buffon Needle problem.

greater than $\frac{1}{2}$ the toothpick crosses more than one line. We determine that $1.25\sin(\theta) = \frac{1}{2}$ in $[0,\pi]$ when $\theta \approx 0.41152$ and $\theta \approx 2.73008$. Thus the probability of the toothpick staying within the lines without crossing any one line, seen in Figure 5.2, is equivalent to

$$2\frac{1}{1.25\pi}\left(\int_0^{0.41152} 1.25\sin(\theta)d\theta + \frac{1}{2}\left(\frac{\pi}{2} - 0.41152\right)\right) \approx 0.348356.$$

∎

Let's look at another problem that is of a much more practical matter.

Example 5.1.2 *Dubey's Pet World has come to you for advice on the following problem. They carry 150 gallon fish tanks which are quite large and take up a lot of shelf space in the store, and thus they do not want more than one on their shelves at a time. They observe that there is not a high demand for these tanks; however, these are a high-profit item and if they don't have one in stock when a customer comes in, the customer will go elsewhere. Moreover, it takes 5 days for a new one to be delivered by the distributor after they place an order. We will use a simulation to determine how many 150 gallon tanks*

Dubey's Pet World sells on average in one month, and how many customers
are turned away, given that a new tank is ordered each time one is sold.

Just as in many modeling problems, we must determine the initial conditions
and the assumptions that will be made in this problem. Given that these tanks
take up a lot of space, we will set the initial number of tanks that Dubey's Pet
World starts with to one tank. We will assume that there can be at most one
potential 150 gallon tank buyer per day.

When discussing how you will run a simulation, it is best to do so in a way
that the reader could go about recreating the simulation. One way you can do
this is with a step by step pseudocode or flowchart. A sample pseudocode can
be seen below with a flow chart for Example 5.1.2 in Figure 5.3 following.

Pseudocode should start with the initial conditions that you are assuming
in the model. For example, we wish to initially set inventory to 1 and the sold
tanks to 0. Keep in mind that the order that steps occur does matter and thus it
is good to articulate in which order the simulation steps are processed, what the
time step unit is, and whether time is discrete or continuous in the simulation.
In the Dubey's Pet World problem, time is discrete and since we are assuming
only one customer comes in per day, the time step that we will use is days as well.

- Step 1: Generate a random binary number to determine if the daily customer
  comes in to buy a tank or not.

- Step 2: If no customer comes to buy a tank today, proceed to the next day.

- Step 3a: If a customer comes in to buy a tank and inventory is 1, then sell
  the tank, set the inventory to 0, increase the number of sold tanks by 1, set
  the number of days until the new tank arrives to 5, and proceed to the next
  day.

- Step 3b: If a customer comes in to buy a tank and inventory is 0, then
  increase the number of customers turned away by 1.

- Step 4a: If inventory is 0 and the count that represents the number of days
  until your order arrives is greater than 0, then decrease this count by 1, and
  proceed to the next day.

- Step 4b: If inventory is 0 and the count that represents the number of days
  until your order arrives is equal to 1, then decrease this count by 1, set the
  inventory to 1, and proceed to the next day.

- Step 5: Repeat Step 1 thirty times (to represent one month).

How many people do you expect Dubey's Pet World to turn away? Should they
be turning away people about half of the time? In fact, although the simulation

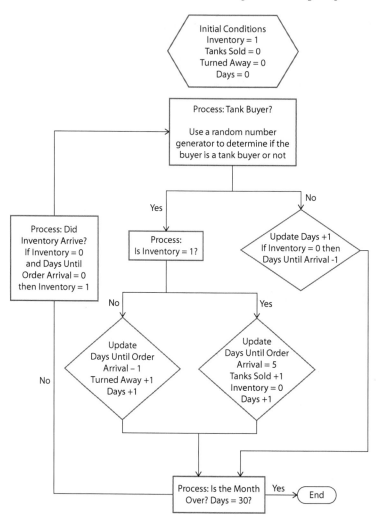

FIGURE 5.3: Flow chart for Dubey's Pet World simulation.

described by the flowchart and pseudocode does tell us what Dubey's Pet World might do during a month, it does not simulate the average tanks sold or average number of customers turned away. In order to simulate an average, we must perform the experiment (or simulation) many times, just as in the Buffon's Needle problem. When you determine what happens at Dubey's Pet World, month after month after month and then take the average, you can produce a histogram similar to the one found in Figure 5.4 to represent what is happening in your simulations.

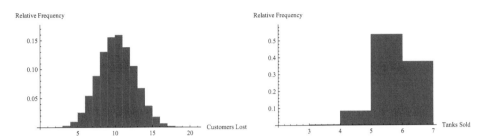

FIGURE 5.4: Histogram of Dubey World Simulation for 50,000 one month runs.

■

These techniques can be used to simulate applications in a variety of fields. Modeling is particularly useful in fields where collecting data is difficult or where ethically initial modeling should proceed physical implementation. For example, if you are in the midst of a disease outbreak and you wish to predict the population that will become infected with the disease in the future, you may not have enough data to work with, but you can use present day behavior of the spread of the disease to model what might happen. In Example 5.1.3, we see an example where scientists may wish to model how an intervention method for a disease affects the population before implementing the intervention method.

Example 5.1.3 *On February 2, 2016, the first Zika virus transmission hit the United States. This particular disease has become a real concern for the western hemisphere, particularly in warmer climates. The Zika virus is transmitted by female Aedes Albopictus mosquitos and this same mosquito transmits another deadly virus called Dengue Fever. In [33], the authors present a possible solution to the transmission of Dengue Fever. In this article, scientists expose the male Aedes Albopictus mosquitos to Wolbachia, which is a natural bacterium that would effectively sterilize them.*

When a male Wolbachia mosquito (MW) mates with a non-Wolbachia female (F), their eggs do not hatch. In addition, Wolbachia females (FW) will always pass on Wolbachia to their offspring.

If we start out with the same number of mosquitos in the each category, MW, FW, F, and M (non-Wolbachia male), how can we construct a simulation to determine the percent of FW mosquitos there are after 8 mosquito gestation periods?

Here are some additional assumptions that we will consider here:

- We will initially start with 6 mosquitos in each category.

- All mosquitos breed at the same time with only one other mosquito.

- If a mosquito has the opportunity to mate then they will.

- All female mosquitos will lay the same number of eggs, 3 male and 3 female. (In real life this number is much larger; however, with a large number we will see the mosquito population grow rapidly.)

- All eggs will mature into breeding mosquitos by the next mating period.

- There is an 80% chance that the mating pair will die after reproducing.

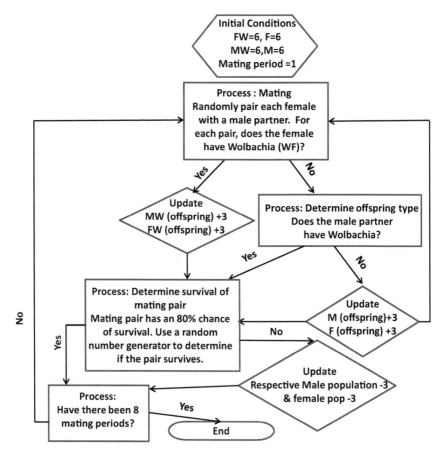

FIGURE 5.5: Flow chart for Wolbachia mosquito simulation.

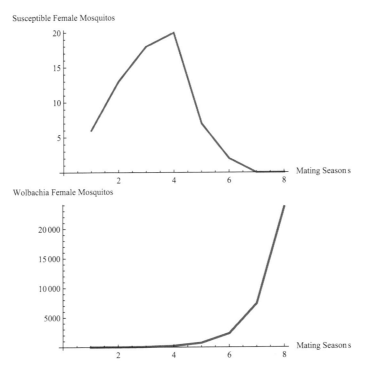

FIGURE 5.6: Possible result of 8 gestation period simulation for Wolbachia spread in mosquitos.

Figure 5.6 shows a diagram of what can happen under a simulation described in Example 5.1.3.

■

Exercises

1. Dubey's Pet World, from Example 5.1.2, has decided to place an order for a new 150 gallon tank each week, instead of after they sell the one that is in stock. Create a new flowchart, altering the one in Figure 5.3, to adjust for this change.

2. ^{TE} Write a simulation to determine, on average, how many tanks Dubey's Pet World will sell, using your flowchart from Exercise 1.

3. The population of elephants in Botswana is growing too rapidly and it is

your job to simulate the sterilization of elephants in your preserve. You know that in any given week you typically see 4 elephants. You are supposed to shoot a sterilization dart at each elephant that you see, but you are not a great shot, and the chance of you hitting any given elephant is .25. Create a flowchart that could be used to describe a simulation to determine how many elephants are sterilized in a one month period. Be sure to include any initial conditions or assumptions that you make as well.

4. ^{TE} Write a simulation to determine, on average, how many elephants you should sterilize in a month, based on the flowchart in Exercise 3.

5. ^{TE} Jack and Ace decide to play a simple betting game based on rolling of a fair sided die. They are each given $100; however, Jack has the advantage because he wins $1 if a 1 or prime number face shows (2,3,5) and Ace wins $1 otherwise. They each decide to play a dollar on each turn. Simulate how much Jack has won after 10 turns.

6. Determine Jack's expected winning after 10 turns in the game outlined in Exercise 5.

7. In Example 5.1.3, discuss how the death rate of the mating mosquitos might affect the final populations in an 8 gestation period cycle.

8. In Example 5.1.3, when a female reproduces, it reproduces 3 female and 3 male offspring. If instead, each time 6 offspring are reproduced with the gender chosen randomly, discuss how this change might affect the final populations in an 8 gestation period cycle.

9. ^{TE} In Section 3.2, Exercise 4 you created a Markov chain model to look at a game of ladder climbing. Write a simulation to determine how long it will take, on average, to finish the game.

Project^{TE}: Trading simulation

There are 400 people in a closed economy with 100 people in each of 4 wealth classes. To start with, each person in the highest wealth class gets a random amount between $4000 and $5000, in the second wealth class each person gets between $2000 and $4000, in the third between $1000 and $2000, and in the last between $0 and $1000. The wealth categories will remain similar to the initial distribution throughout trading, with four categories, greater than

$4000, between \$2000 and \$4000, between \$1000 and \$2000, and less than \$1000 dollars. During each trading period, pairs of traders are chosen at random to trade, where traders can trade within their wealth class or within one wealth class from them. For example, traders in the third wealth class could trade within their wealth class or with the second or first wealth class. When a pair of traders is randomly chosen to trade, they put all of their wealth together and split it in such a way that one trader gets a random amount between 0 dollars and half of the sums of the traders' wealth.

Example: If Trader 1's wealth is \$5200 and Trader 2's wealth is \$3800 then the sum of their wealth is \$9000. Trader 1 is reassigned a random wealth between 0 and 4500 dollars, r, Trader 2 is reassigned $9000 - r$ dollars.

1. Create a flowchart representing the process of trade described. Be sure to include all of your assumptions and initial conditions in your discussion.

2. Create a simulation that will determine the wealth distribution after 10 trading periods in this closed economy.

3. Create a histogram that displays the wealth distribution after 10 trading periods in this closed economy and discuss your findings.

Project^{TE}: Alien Artery Growth Simulation

You have been contacted by a biologist who has discovered an alien species and is trying to describe how the alien's arteries branch as the biologist looks farther away from the alien's heart. The biologist notices that one main artery leaves the heart and after a distance, it branches into 3 separate arteries with one large branch and two smaller "immature" branches. Once a branch is thick enough, it will branch at each fixed distance the same way, but the smaller branches need one additional distance before they branch. The biologist has provided Figure 5.7 to help explain. Your goal is to describe the number of arteries at each branching point. Consider the first "immature" branch artery from the heart to occur at step $n = 0$.

1. Sketch the branching for $n = 4$ and $n = 5$.

2. Describe the number of branches at any step n in terms of previous branching steps. Call the terms in your sequence A_n. (Note: you already know that $A_0 = 1$ and $A_1 = 1$.)

3. Unfortunately, the biologist's alien friend has been in an accident and his hand was severed. The biologist frantically calls you asking how many arteries are in the alien's left hand so he will know how many to fix. Use your

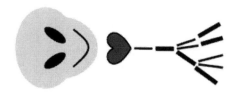

FIGURE 5.7: Example of alien artery growth.

formula from part 2 to find the answer. (Assume that the alien's hand is at the branching step $n = 8$ and that $\frac{1}{3}$ of the branches are in each hand.)

4. Your team now realizes that your alien species grows its arteries based on a space colonization algorithm [34]. The arteries are generated as follows.

   (a) New arteries are generated by attaching a node (the end of an already developed artery) to an influence point (any point within distance $\frac{1}{8}$ inch) with probability .75.

   (b) An influence point can only be attached to one node.

   (c) Once a node is attached to an influence point, the influence point becomes a node.

   Create a simulation using these rules to generate 8 steps of artery growth for this alien.

5.2 Equation-Based Models versus Agent-Based Models

A *difference equation* is a discrete recursive equation for modeling a growth behavior. For example, exponential growth can be represented by a difference equation such as

$$f_{n+1} = 2f_n, \; f_0 = 1.$$

Here $f_0 = 1, f_1 = 2, f_2 = 2^2, \ldots, f_n = 2^n$.

Figure 5.8 shows a discrete plot of the terms represented by the above difference equation model. Notice here that the slope of the secant line through two consecutive points (n, f_n) and $(n + 1, f_{n+1})$ is equal to f_n. When trying to develop a difference equation model, one should try to think about how the rate of change is being altered. In differential equation models we represent this with a derivative, but in the difference equation model, we use the average rate of change. Thus, in this case, we might say that $\frac{f_{n+1} - f_n}{n+1-n} = f_n$ or $f_{n+1} = 2f_n$.

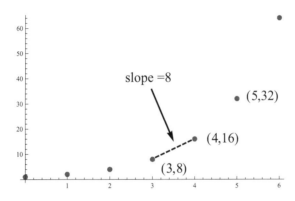

FIGURE 5.8: Plot of terms in difference equation $f_{n+1} = 2f_n, \; f_0 = 1$.

We can look at a similar growth model in a *coupled system* where the interactions between two populations affect each other.

Example 5.2.1 *In a system where there is a predator and a prey, prey grow exponentially without the predator and the predator population decays exponentially without the existence of prey. It is interesting to see how the two populations affect each other in size.*

This behavior, along with the interaction between the predator and prey populations, can be described by the difference equations

$$\begin{aligned} s_{n+1} &= -as_n + bs_n f_n, \\ f_{n+1} &= cf_n - ds_n f_n, \end{aligned}$$

where parameters a, b, c, and d are all positive real numbers, b and d are constants which represent the strength of the influence of the interaction between the two species, and s_i and f_i are the predator and prey population sizes, respectively, at time i. Notice here that the term $bs_n f_n$ represents a gain to the s, or predator, population based on interaction with an f, prey, individual. A predator would potentially gain a survival advantage after encountering prey, and the strength of this increase is quantified through the constant b. Visualize the results of the system for a set of parameters.

We can use this system to attempt to visualize what would happen in the predator prey system when a set of parameters is chosen. For example, if $a = .01, b = .009, c = 1.1, d = .0005$, and $s_0 = f_0 = 100$, then $s_1 = 89, f_1 = 105, s_2 = 83.215, f_2 = 93.2275, s_3 = 68.9892, f_3 = 87.6575, \ldots$. If the populations are plotted, predator versus prey, in time the oscillating behavior can be seen in Figure 5.9.

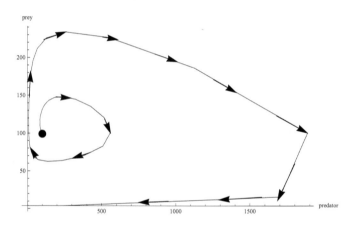

FIGURE 5.9: Predator versus prey using equations from Example 5.2.1.

∎

The predator-prey system can also be modeled with other equations, such as differential equations (see Exercise 4 in Section 4.3).

If we wish to establish a coupled system between two populations, s and f, as in Example 5.2.1, we could think of the rate of change in s as being a function, $p_1(s,f) = -(1 + a)s + bs \cdot f$, and thus

$$\frac{\Delta s}{\Delta t} = p_1(s,f).$$

If we assume that the time step $\Delta t = 1$, then

$$
\begin{aligned}
s_{n+1} - s_n &= p_1(s_n, f_n) \\
s_{n+1} &= s_n - (1+a)s_n + bs_nf_n \\
s_{n+1} &= -as_n + bs_nf_n.
\end{aligned}
$$

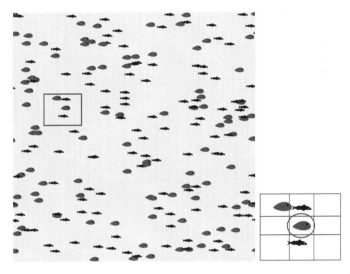

FIGURE 5.10: An example of a predator-prey agent-based model (left) and blow up of grid (right).

Equation-based models usually assume simple interactions with homogeneous mixing and a homogeneous population. With the enhancement of computing power and speed, agent-based models have become more popular. An *agent* is an individual, in a complex system, with its own goals and behavior. Agent-based models usually focus on a discrete number of agents in a grid-like environment, such as the fish and sharks in Figure 5.10, left. In an agent-based model, each agent has its own set of characteristics and terms for which it moves and interacts, adding a level of stochastic behavior into the model. This aspect of modeling allows for heterogeneity among agents; however, this level of detail also requires a larger set of parameters. In addition, most agent-based models integrate a space dimension. The discussion on how to compare equation-based models and agent-based models is a fairly young one [30]. As it should be the goal of a modeler to develop a model which best replicates the natural physical system, one might look at behaviors displayed in agent-based models to be represented in equation-based models.

In the predator-prey agent-based model one might place $f_0 = 100$ prey

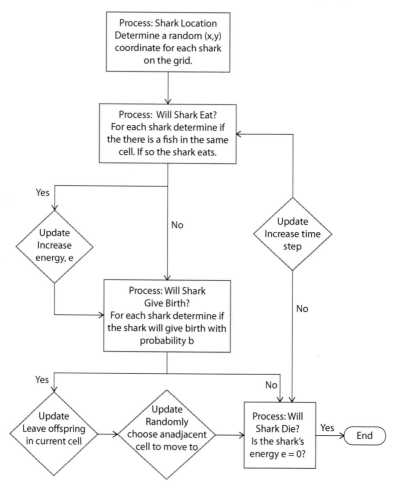

FIGURE 5.11: Flow chart for predator behavior in predator-prey agent-based model.

and $s_0 = 100$ predators randomly on a grid. As we discussed in Section 5.1, it is important to determine your assumptions early in agent-based modeling and further discussion can be done with a descriptive flowchart or pseudocode. Along the same lines of the difference equation model, each agent will have to deal with birth, death, movement, and interaction with agents from the opposite species. Notice in Figure 5.10, right, that the circled fish has a grid around it with 8 adjacent cells around the one it is in. One might choose to move the fish randomly into another cell in the ocean, seen in Figure 5.10, left, or move the fish into one of its 8 adjacent cells. Also when determining the rules for shark and fish interaction in this model, one should determine what will happen to sharks and fish that share the same cell.

One might think about the behavior of the predator (shark) agent using the flowchart in Figure 5.11. Coding the predator-prey agent-based model will be left to the user in the exercises; however, as noted in the discussion of the difference equation predator-prey model, depending on initial conditions, one of three behaviors should be seen: the predator dies out, the prey die out, or the populations show an oscillating behavior.

Agent-based modeling can also be used to simulate other types of biological growth such as tumor growth or to replicate the behavior of a *cellular automaton*. Cellular automata typically involve a grid of cells. Each cell acts under a set of particular rules and can take on a number of states (often represented by different colors). In order to explore cellular automaton with interesting behaviors, we will explore the *Langton's ant* problem.

Example 5.2.2 *Langton's ant, a binary cellular automaton, introduced by Christopher Langton [31], starts out on a grid containing black and white cells, and then follows the following set of rules.*

1. *If the ant is on a black square, it turns right 90 degrees and moves forward one unit.*

2. *If the ant is on a white square, it turns left 90 degrees and moves forward one unit.*

3. *When the ant leaves a square, it inverts the color of that square.*

Figure 5.12 shows the first several hundred steps of Langton's ant through an agent-based model. In fact, Langton's ant appears to be predictable early on but has a fairly complex and unpredictable behavior.

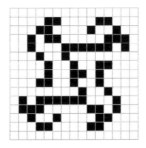

FIGURE 5.12: Agent-based model of Langton's ant over 384 time steps [36].

■

Exercises

1. A community in Sri Lanka currently has 10 female and 10 male pangolin inhabitants. It takes 140 days for a pangolin to give birth to 1 offspring and the offspring is sexually mature at 700 days of age. Each female produces a single offspring every 140 days with a 50% chance of the offspring being female.

   (a) Determine the size of the female pangolin population for $t = 0,1,\ldots,14,15$ where each time period represents 140 days.

   (b) Determine the rate of growth in the female pangolin population in this community between $t = 14$ and $t = 15$ days.

   (c) Let's assume that the pangolin population grows at roughly the same rate as that found in part 1b between any two consecutive birth periods (140 days). Use the rate of growth between $t = 14$ and $t = 15$ as this rate and create a difference equation model for the growth of the female pangolin population in this community.

   (d) Use the model in part 1c to predict the number of pangolins in the community at $t = 16$.

2. ^{TE} The female pangolins from Exercise 1 are antisocial animals and will only choose to reproduce if there is at most one other female pangolin in the same location. Create an agent-based model to determine the number of female pangolins in the community after $t = 2100$ days.

3. In Chapter 5.1 Exercise 3, the goal is to sterilize elephants where 4 elephants are seen per week and the chance of you hitting any given elephant by a sterilization dart is .25. Assume that no elephants are sterilized initially. If e_n represents the population of sterilized elephants in week n, write a difference equation for e_n and plot your results for the first 10 weeks.

4. ^{TE} Assume that there is only one individual shooting sterilization darts at the elephants in Exercise 3 (above) and that individual moves randomly through a grid environment, only shooting an elephant that shares a grid cell with him, and hitting elephants with 25% accuracy.

   (a) Write an agent-based model to model the number of elephants sterilized throughout time, assuming that no elephants are sterilized initially.

    (b) Run your agent-based model for 10 weeks and compare your results for this model with those found in Exercise 3.

    (c) If the individual no longer moves randomly through the grid environment, but instead moves to one of the adjacent cells, choosing a cell with an elephant in it if an adjacent cell contains an elephant. Integrate this movement into your agent-based model and discuss the differences, if any, in your results.

5. Suppose we want to study the population of a certain species of bird. These birds are born in the spring and live at most 3 years, and we will keep track of the population just before breeding. Suppose 20% of Age 0 birds survive to the next spring, and 50% of Age 1 birds survive to become Age 2. Suppose that females that are 1 year old produce a clutch of 4 birds. Females that are 2 years old lay 6 eggs. For $Y0_n$, $Y1_n$ and $Y2_n$, which represent the Age 0, Age 1, and Age 2 birds at year n, determine a system of coupled difference equations.

6. Discuss how an increase in the parameter a in Example 5.2.1 will affect the size of the shark and fish populations.

7. In Figure 5.11, describe how the Process: Will Shark Eat? will directly affect the model behavior of the fish in Example 5.2.1.

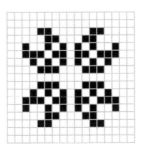

FIGURE 5.13: One state of the pulsar oscillator in the Game of Life.

8. ^{TE} Use agent-based modeling to model the movement of a pulsar oscillator in Conway's Game of Life [28], where a pulsar oscillator starts in the configuration shown in Figure 5.13 and each cell changes its state based on the states of the eight cells around it. Each cell can take on one of two states, life (white) or death (black). When determining the life of a cell, count the

number of cells that are alive (white) surrounding this cell. If the current cell is alive and this count is

(a) less than 2 or greater than 3, the current cell dies and is switched to black,

(b) 2 or 3, the current cell is left unchanged.

If the current cell is not alive and this count is exactly 3, then the current cell is turned on. The current dead cell remains unchanged if the count is anything other than 3.

Project^{TE}: Predator-Prey Agent-Based Model

The goal of this project is to create an agent based model for a predator-prey system. In this model, the initial conditions will be $f_0 = 100$ prey and $s_0 = 100$ predators placed randomly on a grid.

1. Figure 5.11 shows a flowchart for the predator in this system. Create a similar flowchart for the prey in this system.

2. Create an agent-based model with only fish in the system, where no fish die and fish give birth to 10 additional fish, leaving them in their grid cell before they move, at each time step with probability 10%. Run this model for 20 time steps and discuss what happens with the fish population.

3. Add sharks into your agent-based model from part 2, where no shark dies and each shark gives birth to 1 additional shark, leaving them in their grid cell before they move, at each time step with probability 1%.

4. In the agent-based model from part 3, assign each shark an energy level of 5 when they are born. At each time step if a shark shares a cell with a fish they may eat the fish and gain an additional 5 energy points. For each time step that a shark does not eat they lose 1 energy point and if a shark has no energy points then they die. Run this model for 50 time steps and discuss what happens.

5. How sensitive is the model in part 4 to the amount of energy assigned at birth? If the amount of energy assigned at birth is 4 points or 6 points, how does the result of the system behavior change?

Project^{TE}: Agent-Based Models in the Social Sciences

Imagine a grid where each cell is assigned one of three randomly assigned colors, red, green, or white. Each cell will represent a household in a neighborhood, where a white cell represents an empty cell. Each cell can hold at most one household at a time and at each time step households move throughout the neighborhood based on surveying the 8 direct neighbors (cells) around them. If the fraction of neighbors of the opposite color (red and green) directly around a household is greater than a tolerance t, then that household relocates to some other empty cell on the grid.

1. Create a flowchart showing the steps of this agent-based model.

2. Code this agent-based model choosing a set value for t.

3. Determine which values of t will cause your neighborhood to segregate.

Project^{TE}: The Power of Influence

Assign each agent/individual in your model a random real number between -1 and 1 representing the individual's opinion and a random positive real number representing the individual's uncertainty about their opinion. An agent's opinion spectrum is defined by the interval

$$(opinion - uncertainty, opinion + uncertainty).$$

At each time step, each agent interacts randomly with one other agent in the system. If a pair of agents is randomly chosen to interact and their opinion intervals overlap, then they can influence each other's opinions.

If a pair of agents, agent 1 and agent 2, does have overlapping opinion intervals and the uncertainty of agent 1 is greater than the uncertainty of agent 2, then agent 2 will influence agent 1. Under this influence, if agent 1's opinion, $opinion_1$, is less than agent 2's opinion, $opinion_2$, then

$$opinion_1 = opinion_1 + \frac{p(opinion_2 - opinion_1) \cdot (\text{length of overlap in intervals})}{\text{agent 2's uncertainty}}.$$

If $opinion_2 < opinion_1$, then

$$opinion_1 = opinion_1 - \frac{p(opinion_2 - opinion_1) \cdot (\text{length of overlap in intervals})}{\text{agent 2's uncertainty}}.$$

Note that if $opinion_1 = opinion_2$ then neither will influence the other.

1. Create a flowchart showing the steps of this simulation.

2. Those individuals with opinions at or close to 1 or -1 are considered extremists. Create a simulation under the above assumptions about the power of influence, with $p = .5$ and no extremists in your system and determine what happens in the long run.

3. Run your simulation from part 2 with $p = .5$ and several extremists on both ends of the opinion spectrum. Determine what happens to this system in the long run.

4. Determine how sensitive your simulation is to changes in p.

5. Discuss what p might represent in terms of the application.

5.3 Modeling Using Game Theory

Game theory is all about the strategy behind making decisions. As you know, some decisions you can clearly make on your own, some you may have to choose to collaborate, sometimes you may have to change your mind based on others decisions, and sometimes you may choose to make a decision at random.

Let's begin with a look at a famous strategy game called the Prisoner's Dilemma.

Example 5.3.1 *Two partners, Phoenix and Christian, in a crime are arrested and questioned separately. Each criminal has to choose whether or not to confess and implicate the other. If neither man confesses, they each will serve one year in prison. If each confesses and implicates the other, both will go to prison and each serve 10 years. However, if one burglar confesses and implicates the other and the second burglar does not confess, the one who has collaborated with the police will go free, while the other burglar will go to prison for 20 years on the maximum charge. The payoff matrix for the criminals' decision can be seen in Table 5.1.*

TABLE 5.1
Prisoner's Dilemma Payoff Matrix

		Phoenix	
		confess	don't confess
Christian	confess	(10,10)	(0,20)
	don't confess	(20,0)	(1,1)

So what should Phoenix and Christian do in this dilemma?

Notice in Table 5.1 that independent of whether Phoenix confesses or not, it is in Christian's best interest to confess. Similarly, it is in Phoenix's best interest to confess. In this case, we say that the *dominant strategy* is to confess. A dominant strategy occurs if the same strategy is chosen for each of the different combinations of situations the player might face.

■

Although the solution to the Prisoner's Dilemma followed a dominant strategy, it may not have been in the best interest for both players if, for instance, they could have somehow spent just 1 year in jail each. We now explore an example with a *pure strategy*, in which we determine a strategy which is in the best interest of both players.

Example 5.3.2 *Assume that there are two cars stopped at a traffic light traveling at perpendicular directions. Each car can choose to follow traffic laws and* *stop at the light or to violate these laws and proceed through the light. The pay-off matrix for the drivers' decision can be seen in Table 5.2, where $B >> L > 0$.*

TABLE 5.2

Traffic Payoff Matrix

		Car 1	
		STOP	GO
Car 2	STOP	$(-L, -L)$	$(0,L)$
	GO	$(L,0)$	$(-B, -B)$

So what should each car's strategy be at the light?

Notice in Table 5.2 that when both cars are stopped at the light, both drivers lose a small amount of time, L. When one car is stopped at the light and the other proceeds, that car gains that same amount of time; however, if both drivers decide to proceed then they are both at a large loss, B, related to a collision.

A *Nash equilibrium*, named after John Nash, is a strategy in which no player can do better if they individually change their strategy. According to Nash, every game with a finite number of players has a Nash equilibrium. Let's see what strategy in the traffic payoff matrix is a Nash equilibrium.

We will start with the strategy that both cars remain stopped. If car 1 sees that car 2 is stopped then car 1 will go and visa versa; thus this strategy is not an equilibrium. If car 1 sees that car 2 is going through the light it does not benefit car 1 to go and thus this strategy as well as when car 1 goes and car 2 remains stopped are both Nash equilibriums. Clearly the strategy for both cars to go through the light simultaneously is not in the best interest for either driver and thus is not an equilibrum.

∎

Let's look at another game theory example in which it is beneficial to create a payoff matrix. We have all spent a moment in our childhood when we tried to make a decision using Rock-Paper-Scissors. If two players, Bob and Sue, decide to play a game of Rock-Paper-Scissors in which each time the game is played the loser pays the winner 1 dollar, the payoff matrix will look like the matrix in Table 5.3.

TABLE 5.3

Rock-Paper-Scissors Payoff Matrix

		Sue's play		
		Rock	Paper	Scissors
	Rock	$(0,0)$	$(-1,1)$	$(1,-1)$
Bob's play	Paper	$(1,-1)$	$(0,0)$	$(-1,1)$
	Scissors	$(-1,1)$	$(1,-1)$	$(0,0)$

In the Rock-Paper-Scissors game, we know that there is not a dominant strategy; however, this game is a special type of game called a *zero sum game*. This game is called a zero sum game since in each outcome the winner gains what the loser loses and thus there is no net gain between the two players. The goal in most zero sum games is to determine the strategy that will minimize the losses of each player and maximize their profits correlating to each other.

Example 5.3.3 *Sue and Bob decide to play Rock-Paper, where Sue wins 3 dollars if both players show the same object and Bob wins 2 dollars if he shows rock when Sue shows paper and 4 dollars if he shows paper when Sue shows rock. Table 5.4 shows the payoff matrix for this Rock-Paper game. Since the game is a zero sum game, the single number entry represents the gain for Bob and the loss for Sue. (A negative number represents a gain for Sue.)*

TABLE 5.4
Rock-Paper Payoff Matrix

		Sue's play	
		Rock	Paper
Bob's play	Rock	(-3,3)	(2,-2)
	Paper	(4,-4)	(-3,3)

Notice that if Bob plays rock, Sue should play rock to gain 3 dollars. However, if Bob plays paper, Sue should choose to play paper. Thus Sue does not have a dominant strategy. What should Bob's strategy be?

If one strategy benefits Bob, and Bob has no reason to worry about how Sue will play, we say that Bob will play a pure strategy. However, if Bob knew which strategy Sue was playing, he might change his strategy. A *mixed strategy* is one in which the player will play strategy i with probability p_i. It is important to note that the goal, for an individual player, of a mixed strategy is to choose a random strategy that would make the opposing player equally happy with all of their choices.

Bob chooses to play rock with probability p (and paper with probability $(1-p)$), while Sue plays rock with probability q and paper with probability $(1-q)$. If Bob plays rock his expected payoff is $-3q + 2(1-q)$. Similarly, if he chooses paper, his expected payoff is $4q - 3(1-q)$. If Sue decides to mix her strategies, then Bob should not feel like he has to change his choice. Therefore, both payoffs should be the same. This occurs when $-3q + 2(1-q) = 4q - 3(1-q)$. From Figure 5.14, we can see this occurs with probability $q = \frac{5}{12}$. Thus, Sue's mixed strategy should be to choose to play rock $\frac{5}{12}$ of the time and paper $\frac{7}{12}$ of the time.

Bob might also think about a mixed strategy given that when Bob plays paper, Sue would choose paper, while if Bob plays rock, Sue would choose to win

3 dollars with rock. If Sue chooses rock, her expected payoff is $3p-4(1-p)$ while paper will yield an expected payoff of $-2p + 3(1 - p)$; see Figure 5.14. Again, this means Sue would be equally happy with either payoff if $3p - 4(1 - p) = -2p + 3(1 - p)$. Thus, Bob's mixed strategy is to play rock $\frac{7}{12}$ of the time and paper $\frac{5}{12}$ of the time.

In the case where both players, Bob and Sue, are playing a mixed strategy where both opposing players would be happy with either of their choices, and thus have no reason to make a change, we say that we have a *mixed-strategy equilibrium*.

Using this information, what is the probability that Bob will win 2 dollars

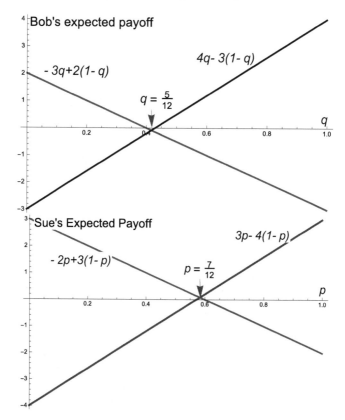

FIGURE 5.14: (Top) Bob's expected payoff. (Bottom) Sue's expected payoff.

and Sue will lose 2 dollars?

$$P(\text{Both players choose paper}) = \frac{7}{12} \cdot \frac{5}{12},$$
$$P(\text{Both players choose rock}) = \frac{7}{12} \cdot \frac{5}{12},$$
$$P(\text{Bob chooses rock and Sue chooses paper}) = \frac{7}{12} \cdot \frac{7}{12},$$
$$P(\text{Bob chooses paper and Sue chooses rock}) = \frac{5}{12} \cdot \frac{5}{12}.$$

We can also see that the expected payoff for Bob is $-3 \cdot \frac{7}{12} \cdot \frac{5}{12} + 2 \cdot \frac{7}{12} \cdot \frac{7}{12} + 4 \cdot \frac{5}{12} \cdot \frac{5}{12} - 3 \cdot \frac{5}{12} \cdot \frac{7}{12} = \frac{-1}{12}$.

■

Exercises

1. You and your friend have decided to go out for a movie. You prefer to see a comedy but your friend prefers to see an action thriller. If you and your friend go to the comedy then you will get to choose what to do for 3 nights this week and Bob gets to choose what to do for 2 nights this week. If you both go to the action thriller then your friend gets to choose what to do 3 nights this week and you get to choose 2 nights. If you go to the comedy and your friend goes to the action thriller then you both get to choose what to do for 1 night this week. If you go to the action thriller and your friend goes to the comedy then there is no payoff for either of you.

   (a) Construct the payoff matrix for this decision.

   (b) Determine the mixed strategy equilibrium for this game.

2. A penalty kick in soccer is between two players, the goalkeeper and the shooter. While the shooter can choose to shoot to the left or right of the goalkeeper, the goalkeeper must decide which direction to dive. If the goalkeeper guesses incorrectly then a goal is scored; otherwise we will assume that no goal is scored. Construct a payoff matrix, assuming a zero-sum game, and determine if there is a pure or mixed strategy; if an equilibrium exists find it.

3. In the penalty kick decision from Exercise 2, explain why the kicker does not have a dominant strategy.

4. Going into the second period of his wrestling match, Johan is trying to decide whether to attack or wait for Lee to attack, Lee is leading the match at this point. If Johan attacks, catching Lee off guard, then he will have

a 60% chance of beating Lee, which he values as much as a point win, 15 points. If Johan attacks, there is a 40% probability that Lee will escape and win the match. If Lee attacks, catching Johan off guard, he will have a 40% chance of winning. Lee too values this to be a 15 point win. If both wrestlers attack at the same time, Lee has an 80% chance of winning. In either case, a loss is worth 0 points. The next person to score a point of any value will win the match; however, if no one attacks the score will remain the same.

(a) Determine Johan's expected value given that he attacked and Lee waited.

(b) Determine Lee's expected value given that Johan attacked and he waited.

(c) Determine Johan's expected value given that both he and Lee attacked simultaneously.

(d) Construct the payoff matrix for this decision.

5. The payoff matrix for a two person decision can be seen in Table 5.5.

TABLE 5.5

		Player A	
		Choice 1	Choice 2
Player B	Choice 1	(4,-4)	(-2,2)
	Choice 2	(-1,1)	(3,-3)

(a) Determine Player A's mixed strategy.

(b) Determine Player B's mixed strategy.

6. A group of students are going on a service learning trip on which they will live close together. Where they are going, there is a disease which spreads easily among people who live close together. The value of the trip to a student who does not get the disease is 6. The value of the trip to a student who gets the disease is 0. There is a vaccination, with a 100% success rate, against the disease. The vaccination costs different amounts for different students. This year, the group is a small group of 3 students. In order to keep track of the students, they are each assigned a unique student number, 1 through 3. The vaccination costs a value of i for the student with student number i, where $1 \leq i \leq 3$. If a student is not vaccinated the probability of getting the disease equal to $\frac{\text{The number of unvaccinated students}}{\text{Total number of students}}$.

(a) Determine student i's payoff if they get the vaccine, for $i = 1,2,3$.

(b) Determine student i's payoff if they do not get the vaccine and one other student gets the vaccine.

(c) Determine student i's payoff if they do not get the vaccine and both of the other students get the vaccine.

(d) Determine how your answer would change in part 6b if there were 4 students in the group.

(e) Determine student i's payoff if there are 4 students on the trip and student i does not get the vaccine but all of the other students get the vaccine.

7. Two players, A and B, are playing a simple coin game. Both players are given a fair sided coin. If both players show heads then player A gets \$3 and if they both show tails then player A gets \$1. If players A and B play opposite faces then player B wins \$2.

(a) Create a payoff matrix.

(b) Determine the mixed strategy for players A and B.

(c) Determine the expected winnings of player A and the expected winnings of player B if this game is played just once and if the two players repeat this process 16 times.

8. Two bars charge their own price for a beer, either 2, 4, or 5 dollars. The cost of obtaining and serving the beer can be neglected. It is expected that 6000 beers per month are drunk in a bar by tourists, who choose one of the two bars randomly, and 4000 beers per month are drunk by natives who go to the bar with the lowest price, and split evenly in case both bars offer the same price. Create the payoff matrix for this problem.

Project^{TE}: Conflict Resolution

Game theory can be particularly interesting when it comes to analyzing conflict resolution. In this project, we will consider players involved in an ethnic conflict. Note that these two players in fact will each represent a group of players with the same intentions. In [32], the authors present a cumbersome model of such conflicts in which each player has two possible strategies, cooperate or defect, where the payoff matrix can be seen in Table 5.6.

TABLE 5.6

Payoff Matrix for Conflict Resolution

		Player 2	
		cooperate	defect
Player 1	cooperate	(α_1,α_2)	(a_1,b_2)
	defect	(b_1,a_2)	(γ_1,γ_2)

The parameter b_i represents the ith player's incentive for defection.

1. Let $a_1 = a_2 = 2, \alpha_1 = \alpha_2 = 1$, and $\gamma_1 = \gamma_2 = .1$. Note that smaller groups are much more likely to defect than larger groups. If n_1 and n_2 are the sizes of our player's groups, and $n_1 > n_2$, determine how b_1 and b_2 are related.

2. Construct an expected payoff function for each of the players if the opponent chooses to cooperate, using the values from part 1. Note that the expected payoff function for group i will be a function of b_i.

3. Determine player 1's mixed strategy for several values of b_1 where $1.1 < b_1 < 1.9$.

4. Recall that b_1 is player 1's payoff for defecting if player 2 cooperates. Let's assume that as time goes on, the number of negotiation periods or time in the negotiation period increases, more if player 1's people join the movement to cooperate and b_1 decreases linearly by some small amount, ϵ, and α_1 increases linearly by that same amount. Create a new expected payoff function for player 1, under the assumption that player 2 will cooperate, with b_1 and α_1 as a function of time. Use this function to determine the probability that player 1 cooperates after 8 negotiation periods.

5. Now let's assume that b_1 decreases exponentially in time by some small rate, ϵ, and α_1 increases exponentially by that same amount. Create a new expected payoff function for player 1, under the assumption that player 2 will cooperate, with b_1 and α_1 as a function of time.

6. If $b_1 = 1.1$ and $\epsilon = .01$ in the expected payoff function from part 5, determine when player 1 is least likely to cooperate.

Project: Braess' Paradox (Another Traffic Model) [26]

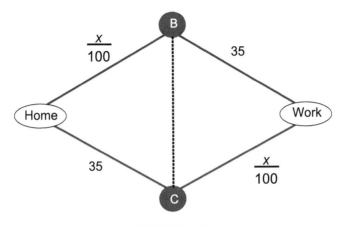

FIGURE 5.15

Three thousand members of a community travel from the same home town to the same work location, in separate vehicles, every morning. There are two different routes to go from home to work, shown in Figure 5.15, one through point B and one through point C (the dotted route is still under construction and not open to vehicles yet). To go from home to point C, and from point B to work, it takes 35 minutes. The time that it takes to travel from home to point B is dependent on the number of vehicles (x) on the road. This is also true about the time it takes to travel from point C to work.

1. If half of the community choose to go on each route, how long will it take for every one to go from home to work?

2. Is this the best strategy for all of the commuters? (Determine if the strategy in part 1 is a Nash equilibrium.)

3. Now let's assume that there is a very short path, of essentially 0 time to travel, between points B and C. If all vehicles travel from home to point B, from point B to point C, and then on to work from point C, how long will it take them to make the journey?

4. Explain why the strategy in part 3 is a Nash equilibrium. What we see happen here is that by building an alternate route, we have caused the journey to take longer. This result is because all the drivers are acting out of self-interest and not as a community.

5.4   Optimization with Voronoi Diagrams

In calculus, we have many problems where we wish to optimize some form of distance. In this section, we will discuss how to optimally divide a plane into sections based on the distance from a set of points (called *seed points*). Our search is for a set of points, for each seed point, that is closest to that seed. We will use a *Delaunay Triangulation* to find these sets and visually represent the results of this algorithm with a *Voronoi Diagram*.

If there are only two seed points, b_1 and b_2, in a plane, then the problem of dividing the plane is straightforward. Draw a line segment between b_1 and b_2 and then draw the perpendicular bisector to that line segment. This perpendicular bisector will divide the plane into two *cells*; all of the points in one cell are closest to b_1 and all of the points in the other are closest to b_2. The points on the perpendicular bisector are equidistant from each seed point.

With a larger number of seed points, the problem becomes more complex. In order to solve this problem, we will introduce the construction of a Voronoi diagram.

Example 5.4.1 *The local municipalities have decided to restructure the voting districts based on the residents' distance from the nearest municipal building. The map of the local municipal buildings can be seen in Figure 5.16. They wish to use a Voronoi diagram to do so.*

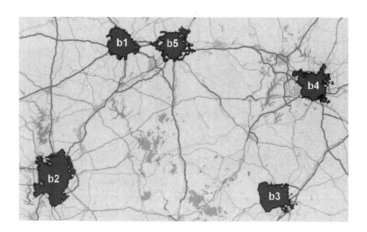

FIGURE 5.16

We begin with an example of a Voronoi diagram with three buildings and will expand the algorithm to a larger set of buildings afterward. Let's begin our exploration with municipal buildings b_1, b_2, and b_3.

To construct the Voronoi diagram with three points,

1. Create a circle that goes through all three seed points b_1, b_2, and b_3.

2. Mark the center of the circle from Step 1.

3. Draw line segments between each pair of seed points.

4. For each line segment from Step 3, draw a ray from the center of the circle, from Step 2, that is perpendicular to and bisects that line segment.

The three rays that you create form the Voronoi diagram where all of the points in a cell are closest to the seed point in that cell.

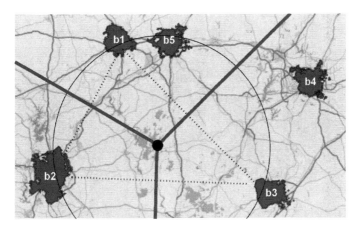

FIGURE 5.17: Example of Voronoi diagram with 3 seed points.

Expanding this algorithm for a set of seed points greater than 3, we will use an algorithm called the Delaunay triangulation.

1. For every set of three noncollinear seed points, create a triangle with the seeds as the vertices. Notice that if three seeds are collinear, lying along a straight line, then there is no triangle to create.

2. Create the circumcircle of the triangle, a circle in which this triangle, created in Step 1, is circumscribed. Mark the center of this circle; this point is called the *circumcenter* of the triangle. If this circumcircle includes any seed points in its interior then do not include this triangle in the Delaunay triangulation. Once you complete this step delete the circle created. An example of when this occurs can be seen in Figure 5.18 (Top) and an example of a Delaunay triangulation in Figure 5.18 (Bottom).

The Voronoi diagram is called the *dual* graph of the Delaunay triangulation.

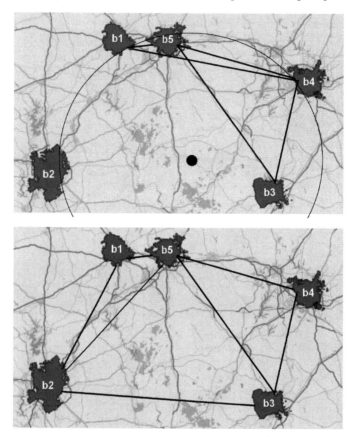

FIGURE 5.18: Example of Delaunay triangulation.

More specifically,

- The vertices of the Voronoi diagram are the circumcenters of the triangles in the Delaunay triangulation.

- The edges of the Voronoi diagram are the perpendicular bisectors of the edges in the Delaunay triangulation from the respective circumcenters.

∎

Voronoi diagrams and Delaunay triangulations are used for modeling in a variety of fields, including protein folding [27], modeling terrains, robotics, GIS, and simulating brain tumors [29].

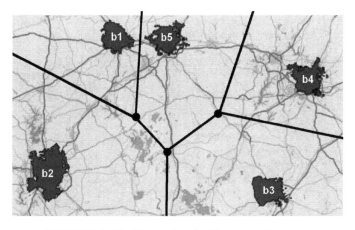

FIGURE 5.19: Example of a Voronoi diagram.

Exercises

1. Discuss why it might be of interest to a pilot to have a Voronoi diagram where each of the seed points are airports.

2. (a) Find the circumcircle and circumcenter for the triangle shown in Figure 5.20.

FIGURE 5.20

(b) Find the Delaunay triangulations for the 4 seed points in Figure 5.20.

(c) Given the four points in Figure 5.20, find the Voronoi diagram.

3. (a) Assume that a forest fire will spread throughout the Smokey National Forest starting from the 3 source points shown in Figure 5.21. Draw

the Voronoi diagram to determine which points will most likely be engulfed by each of the source fires.

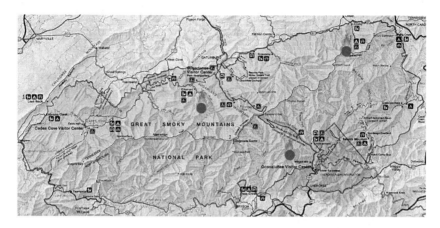

FIGURE 5.21

(b) Discuss why it might be helpful for the national park to start their fire fighting efforts from the lines in the Voronoi diagram.

4. Figure 5.22 shows a representation of normal cells which typically take on an orderly cell arrangement. Typically cancer cells take on polygonal shapes of unequal sizes.

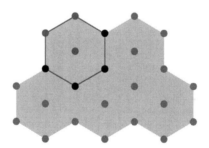

FIGURE 5.22: Representation of normal cells.

(a) Create 3 Voronoi cells using the 3 red points and the red border in Figure 5.22, changing the upper left cell into 3 cancerous cells.

(b) Create 4 Voronoi cells using the 4 black points and the red border in Figure 5.22, changing the upper left cell into 4 cancerous cells.

Project^{TE}: Delaunay Triangulation and Terrain Modeling

The local terrain looks like the function

$$f(x,y) = \frac{4}{4 + (x+5)^2 + y^2} + \frac{4}{2 + (x-2)^2 + (y-5)^2} + \frac{4}{2 + (x-5)^2 + (y-5)^2},$$

where $-10 \le x \le 10, -10 \le y \le 10$.

1. Plot the function above that describes the terrain.

2. Choose 10 data points on this terrain curve that you believe best display the behavior of the curve.

3. Create a Delaunay triangulation using these 10 points to create a model of the terrain.

Project: Voronoi Fractals

1. Choose 3 to 5 randomly selected points, seed points, across the plane and create their Voronoi cells.

2. In each of the Voronoi cells, choose 3 to 5 randomly selected seeds and create the Voronoi diagram in each of the cells from the Step 1.

3. Repeat Step 2, 1 to 3 more times.

4. Color your regions as desired; an example can be see in Figure 5.23.

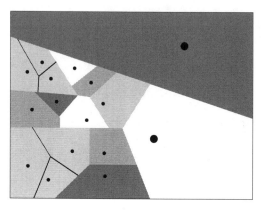

FIGURE 5.23: Example of beginning steps of a Voronoi fractal.

Chapter Synthesis: Independent Investigation

1. (Revisiting Exercise 3 from Section 5.1 and Exercise 3 from Section 5.2)
   As you were informed in a previous section, the population of elephants in
   Botswana is growing too rapidly and it is your job to simulate the sterilization of elephants in your preserve. You know that in any given week you
   typically see 4 elephants. You are supposed to shoot a sterilization dart at
   each elephant that you see.

   (a) Consider a population within the preserve of 100 elephants. Assume
       that no elephants are sterilized initially, and your chance of hitting
       any given elephant is 0.25. If e_n represents the population of sterilized
       elephants in week n, write a difference equation for e_n.

   (b) You are not a great shot, but you could go through training that would
       increase your accuracy to 0.4, 0.5, or 0.6 (at different costs). Write dif-
       ference equations as in part 1a for each of these accuracies. Plot your
       results for all four difference equations for the first 10 weeks on the
       same set of axes.

   (c) Your boss would like your input on how to proceed. You want to help
       control the population, but you do not want to sterilize all the ele-
       phants. The government is pressuring your boss to have the population
       under control in the next 10 weeks. Your boss wants you to develop a
       strategy that the preserve can afford, which can be one of the following:

       - Training for you to increase your shooting accuracy to 0.5,

- Extra support (gas, vehicles) so that you can go around more of the preserve and likely encounter 6 elephants per week, or
- Training for you to increase your shooting accuracy to 0.6 but with reduced support for your traveling around the preserve so that you probably only encounter 2 elephants per week.

You need to have at least 10% of the population sterilized in the next 10 weeks, but if you sterilize a higher percentage, you risk taking a negative toll on the population's survival. Create a report for your boss that includes the following:

i. A plan using one of the three options above including a suggested timeline for the sterilization.

ii. Suggestions on which plan to use based on different cost scenarios, such as if training is more expensive than preserve resources or vice versa.

2. ^{TE}**Modeling a Zombie Apocalypse**

(a) Zombie warlords have taken over the city of Encrypt and they are fighting over territory. These 4 warlords must decide on the boundaries of their territory. The zombie warlords are located in black on Figure 5.24; use a Voronoi diagram to determine each warlord's territory.

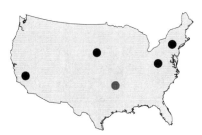

FIGURE 5.24

(b) If the red dot on Figure 5.24 is the only human in the area, discuss the human's best plan to survive as long as possible.

(c) A zombie can bite, and infect, two people on a given day; however, zombies are much slower than humans and thus have a low probability, p_1, of capturing and infecting a human when they are not in a mob. If you start with z_0 zombies and h_0 humans, create a difference equation model for the zombie and human populations.

(d) Create a flowchart that would be useful in describing a model of how the zombie population would grow in the town of Encrypt under the circumstances described in part 2c.

(e) (technology heavy) Create an agent-based model for the zombie outbreak using the flowchart from part 2c. Choose p_0, z_0, and h_0 and discuss how the outbreak is affected by changes in these values.

(f) (technology heavy) If zombies are in a mob, more than 10 zombies in close proximity, within a radius r, the probability of human capture is much higher. Since zombies work best in a mob, an individual zombie will choose to move toward another zombie and once a mob is created the mob will move toward humans together. Incorporate this feature into your model in part 2e and discuss the speed of the zombie outbreak and how the parameter r affects this outbreak.

(g) Randomly place 4 zombie warlords in the agent-based model from part 2f. A zombie morphs into a zombie warlord if it bites and infects at least one human 3 days in a row. In addition, once a zombie becomes a warlord, that warlord will not move within radius r of another warlord and in fact will attempt to move away from any other warlord. Discuss how the incorporation of zombie warlords affects the growth of the zombie population in Encrypt.

(h) A human (in red) is faced with one zombie (in black) in front of them and a mob of zombies (in gray) behind them, as in Figure 5.25. The human, zombie, and zombie mob can only move 1 adjacent grid space per time step, or stay in the same place. Use game theory to argue as to which direction the human should move.

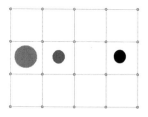

FIGURE 5.25

(i) Discuss any further game theory type encounters the humans, zombies, and zombie warlords might have and what the strategies for the involved parties might be.

Bibliography

[1] J. Enders, S. Kas-Danouche, W. Liao, B. Rasmussen, T. Anh Vo, K. Yokley, L. Robertson, and R.C. Smith (2003). Design of a Membrane Deployable Structure. Proceedings of the 44th AIAA/ASME/ASCE/AHS Structures, Structural Dynamics, and Materials Conference, Norfolk, VA, AIAA-2003-1494. (Also see CRSC Technical Report, CRSC-TR03-06 http://www.ncsu.edu/crsc/reports/ftp/pdf/crsc-tr03-06.pdf

[2] K.D. Monahan (2007). Effect of aging on baroreflex function in humans. *American Journal of Physiology: Regulatory, Integrative and Comparative Physiology*, 293(1), pp. R3-R12, doi: 10.1152/ajpregu.00031.2007

[3] R.F. Rea and D.L. Eckberg (1987). Carotid baroreceptor-muscle sympathetic relation in humans. *American Journal of Physiology: Regulatory, Integrative and Comparative Physiology*, 253(6), pp. R929-R934.

[4] "Greensboro, NC, North Carolina, USA: Climate, Global Warming, and Daylight Charts and Data." http://www.climate-charts.com/Locations/u/US72317003136301.php (Recorded in July 2016.)

[5] www.drawinghowtodraw.com

[6] E.W. Weisstein, "Circumcenter." From MathWorld–A Wolfram Web Resource. http://mathworld.wolfram.com/Circumcenter.html

[7] "Math of drugs and bodies (pharmacokinetics)." http://www.intmath.com/blog/mathematics/math-of-drugs-and-bodies-pharmacokinetics-4098 (Recorded in July 2016.)

[8] C. Arangala and N. Mistry (2015). Music genomics: Applying seriation algorithms to Billboard #1 hits. *Discrete Mathematics, Algorithms and Applications*, Vol. 7, No. 4 (26 pages).

[9] C. Arangala, J.T. Lee, and C. Borden (2014). Seriation algorithms for determining the evolution of the Star Husband Tale. *Involve*, Vol. 7. Number 1, pp. 1-14.

[10] J.E. Atkins, E.G. Boman, and B. Hendrickson (1999). A spectral algorithm for seriation and the consecutive ones problem, *SIAM J. Comput.* 28:1. pp. 297–310. MR 99j:68049 Zbl 0930.05064

[11] Kurt Bryan and Tanya Leise (2006). The $25,000,000,000 eigenvector: The linear algebra behind Google. *SIAM Rev.*48, no. 3, 569–581 (electronic). MR 2278443 (2008b:15030)

[12] Centers for Disease Control and Prevention, http://www.cdc.gov/zika/geo/united-states.html. Last viewed March 16, 2017.

[13] X. Dong, B. Milholland, and J. Vijg (2016). Evidence for a limit to human lifespan. *Nature*, Vol. 538, No. 7624, pp. 257-259.

[14] A. Dundes (1965). *The Study of Folklore.* Englewood Cliffs, Prentice-Hall Inc.

[15] Mouse Gene Expression at the BC Cancer Agency, http://www.mouseatlas.org/. Last viewed March 24, 2017.

[16] National Weather Service. US States and territories. Silver Spring (MD): US Department of Commerce; 1999, www.nws.noaa.gov/geodata/catalog/national/html/us_state.htm. Last viewed March 18, 2017.

[17] R.W. Pollay. (1983). Measuring the cultural values manifest in advertising. *Current Issues and Research in Advertising*, 6(1), pp. 71-92. doi: 10.1080/01633392.1983.10505333.

[18] A. Shuchat (1984). Matrix and network models in archaeology, *Math. Mag.* 57:1. pp. 3–14. MR 86a:00017 Zbl 0532.90097

[19] Matthew Simonson (2016). Hookups, Dating, and Markov Chains: Teaching Matrices so that Your Students Won't Hate Them, Part III. *AMS Blogs*, https://blogs.ams.org/mathgradblog/2016/03/02/markov-chains/. Last viewed October 31, 2017.

[20] United States Census Bureau, Population Estimates. http://www.census.gov/popest/data/state/totals/2015/index.html. Last viewed March 16, 2017.

[21] U.S. Department of Education, Fiscal Years 2015-2017 State Tables for the U.S. Department of Education, http://www2.ed.gov/about/overview/budget/statetables/17stbystate.pdf. Last viewed March 16, 2017.

[22] U.S. Department of Education, New State-by-State College Attainment Numbers Show Progress Toward 2020 Goal, http://www.ed.gov/news/press-releases/new-state-state-college-attainment-numbers-show-progress-toward-2020-goal. Last viewed March 16, 2017.

[23] United States - Life Expectancy at Birth, http://www.indexmundi.com/facts/united-states/life-expectancy-at-birth. Last viewed March 16, 2017.

[24] http://www.historynet.com/battle-of-gettysburg

[25] D. Daugherty, T. Roque-Urrea, J. Urrea-Roque, J. Troyer, S. Wirkus, and M.A. Porter (2009). Mathematical models of bipolar disorder". *Commun Nonlinear Sci Numer Simulat*, 14, pp. 2897-2908.

[26] W. Chen (2016). Bad Traffic? Blame Braess' Paradox. *Forbes*. https://www.forbes.com/sites/quora/2016/10/20/bad-traffic-blame-braess-paradox/#4119b0d114b5. Last viewed March 14, 2017.

[27] F. Dupuis, J-F. Sadoc, R. Jullien, B. Angelov and JP. Mornon (2005). Voro3D: 3D Voronoi Tessellations Applied to Protein Structures. *Bioinformatics*, 21(8), pp. 1715-1716.

[28] M. Gardner. The Game of Life, Parts I-III. Chs. 20-22 in *Wheels, Life, and Other Mathematical Amusements*. New York: W. H. Freeman, 1983.

[29] A.R. Kansal, S. Torquato, G.R. Harsh IV, E.A. Chiocca, and T.S. Deisboeck (2000). Simulated brain tumor growth dynamics using a three-dimensional cellular automaton, *Journal of Theoretical Biology*, 203:4, pp. 367-382.

[30] J.S. Koopman, G. Jacquez, and S.E. Chick (2001). New data and tools for integrating discrete and continuous population modeling strategies. *Annals of the New York Academy of Sciences*, 954(1), pp. 268-294.

[31] C.G. Langton. (1986). Studying Artificial Life with Cellular Automata. *Physica*, 22D, pp. 120-149.

[32] L. Luo, N. Chakraborty, and K. Sycara (2009). Modeling ethno-religious conflicts as prisoner's dilemma game in graphs. In *Computational Science and Engineering, 2009. CSE'09*. vol. 4, pp. 442-449.

[33] S. O'Neill (2015). How a tiny bacterium called wolbachia could defeat dengue, *Scientific American*, June.

[34] A. Runion, B. Lane, and P. Prusinkiewicz (2007). Modeling trees with a space colonization algorithm. *Eurographics Workshop on Natural Phenomena.* pp. 63-70.

[35] A. Rezhdo, Buffon's Needle Problem. https://math.dartmouth.edu/~m20f11/RezhdoProj.pdf. Last viewed March 14, 2017.

[36] Wolfram MathWorld. Langton's Ant. http://mathworld.wolfram.com/LangtonsAnt.html. Last viewed March 14, 2017.

Solutions

1.1 Solutions

1. The company needs to produce more than 7 products.

2. (a) After 1 hour, the number of bacteria will double to 100. After the next hour, the bacteria will double again and be 200.

   (b) In the previous part of the problem, you were assuming that the bacteria had unlimited resources and space, which may not actually be true.

3. (a) $A(12) = 6.25 < A(10) < 12.5 = A(9)$

   (b) 9.92 g

4. Eleven years after the first toy was available for sale, the function predicts $191,583 profit. After 12 years, the function predicts $280,000 profit. The prediction after 11 years is probably more accurate because the model is representative of the first 10 years.

5. (a) 25 mg/h

   (b) 21.28 mg/h

   (c) 7 mg/L

6. A bull that is between 4 and 5 years old because he has a higher probability of fertility.

1.2 Solutions

1. (a) See figure below.

    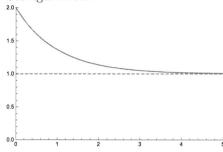

    (b) See figure below.

    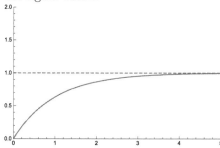

    (c) See figure below.

    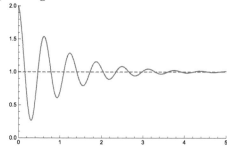

2. The function $f(x)$ is increasing on about $(2.7, 4)$ and decreasing on about $(-\infty, 2.7) \cup (4, \infty)$. The critical numbers of $f(x)$ are about 2.7 and 4. $f(x)$ has a relative maximum at about $x = 4$ and has a relative minimum at about $x = 2.7$. The function $f(x)$ appears to be concave up on about $(2.2, 3.4) \cup (4.5, \infty)$ and concave down on about $(-\infty, 2.2) \cup (3.4, 4.5)$ which means the function has approximate inflection points at $x = 2.2$, $x = 3.4$, and $x = 4.5$.

The end behavior can be described by

$$\lim_{x \to \infty} f(x) = 0$$

$$\lim_{x \to -\infty} f(x) = 0$$

which means that the function outputs approach the horizontal asymptote $y = 0$ as $t \to -\infty$ and as $t \to \infty$.

3. (a) The function $f(x)$ is increasing on about $(-\infty,0)$ and decreasing on about $(0,\infty)$. $f(x)$ has one critical number, 0. $f(x)$ has a relative maximum at about $x = 0$. The function $f(x)$ appears to be concave up on about $(-\infty, -1) \cup (1,\infty)$ and concave down on about $(-1,1)$ which means the function has approximate inflection points at $x = -1$ and $x = 1$. The end behavior can be described by

$$\lim_{x \to \infty} f(x) = 1$$

$$\lim_{x \to -\infty} f(x) = 1.5$$

which means that the function outputs approach 1 as $t \to \infty$ and 1.5 as $t \to -\infty$.

(b) The maximum value is about 7.

4. (a) Six years after the first product is available.

(b) Between the end of year 6 and the beginning of year 10.

(c) At the end of year 8.

5. (a) As $x \to \infty$, $B(x)$ approaches a horizontal asymptote $B = 100$. The number of bacteria in the petri dish is limited to 100.

(b) $B(x)$ is always increasing and never decreasing. The number of bacteria is always getting larger.

(c) Around $x = 4$.

6. (a) Age of approximately 4.8 years.

(b) Age of approximately 3.4 years.

(c) Answers may vary. One recommendation would be to only bring in bull elephants younger than 6 years of age.

7. $C = 100$ and $k = 80$.

1.3 Solutions

1. (a) $g(x) = (x - 2)^2 + 1$

(b) The relative minimum is moved to the right 2 units and up 1 unit.

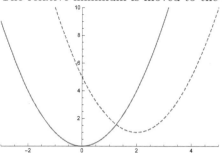

2. The two functions are the same. Reflecting $f(x) = |x|$ about the y-axis creates an identical graph because $f(x) = |x|$ is symmetric about the y-axis.

3. (a) Answers may vary. $g(x)$ appears to be transformed from moving $f(x) = \sqrt{x}$ to the right 5 units before flipping the graph about the y-axis. Then the function is multiplied by a factor greater than 1.

   (b) $g(x)$ is always decreasing, but $f(x)$ is always increasing. Both curves appear to always be concave down. $f(x) \to \infty$ as $t \to \infty$, but $g(x) \to \infty$ as $t \to -\infty$.

   (c) $g(x) = 2\sqrt{5 - x}$

4. (a) Answers may vary. $g(x)$ appears to be transformed from moving $f(x) = x^3$ to the right 1 unit and multiplying it by a factor larger than 1. The graph has also been moved down about 4 units.

   (b) Both curves are always increasing. Both $f(x)$ and $g(x)$ approach ∞ as $x \to \infty$ and approach $-\infty$ as $x \to -\infty$. Neither function has a relative maximum nor a relative minimum. Each curve has one inflection point. $f(x) = x^3$ has one inflection point at $x = 0$, and $g(x)$ has one inflection point at about $x = 1$. Both curves change from being concave down to concave up at their respective inflection points.

   (c) $g(x) = 10(x - 1)^3 - 4$

5. (a) Answers may vary. $g(x)$ appears to be transformed by moving the graph to the left about $1/2$ unit before reflecting the the curve about the y-axis. Then the function is multiplied by a factor $-a$ where a is a number between zero and 1.

   (b) Both curves have 3 extrema and 2 inflection points. $f(x)$ decreases on $(-\infty, -1)$, increases on $(-1,0)$, decreases on $(0,1)$, and increases on $(1,\infty)$. $g(x)$ increases on $(-\infty, -\frac{3}{2})$, decreases on $(-\frac{3}{2}, -\frac{1}{2})$, increases on $(-\frac{1}{2}, \frac{1}{2})$, and decreases on $(\frac{1}{2}, \infty)$. When $f(x)$ has relative extrema at $x = a$, $g(x)$ has the opposite type of extrema at $a - \frac{1}{2}$. Similarly, $f(x)$ is concave up on about $(-\infty, -0.6)$ and on about $(0.6, \infty)$ (and concave down otherwise) while $g(x)$ is concave down on about $(-\infty, -1.1)$ and on about $(0.1,\infty)$ (and concave up otherwise). The inflection points for $f(x)$ occur at about $x = -0.6$ and $x = 0.6$, and the inflection points for $g(x)$ occur at about $x = -1.1$ and $x = 0.1$.

(c) $g(x) = -\left(x - \frac{1}{2}\right)^2 \left(x + \frac{3}{2}\right)^2 + 2$

6. (a) $q(t) = \frac{72000}{(t-1)^{\frac{3}{2}}} e^{-\left(\frac{30}{t-1} + t - 3\right)}$

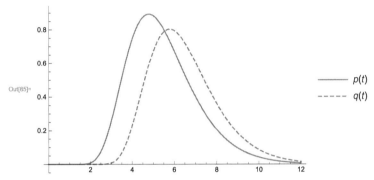

(b) If a bull is available that is greater than 5.5 years old, the zoo would hope the bull is on the new drug. If the zoo finds bulls that are younger than 5.5 years to bring in, the zoo would like them not to have been on the drug.

1.4 Solutions

1. (a) m has a larger effect because $f'(x) = m$.

    (b) b has a larger effect because $f(0)$ does not depend on m at all.

2. (a) $A'(t) = 2000re^{rt}$

    (b) See figure below.

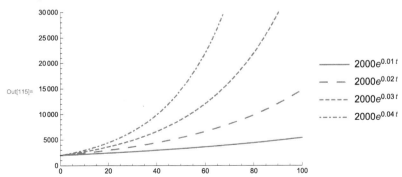

(c) Answers may vary. $A(t)$ increases more rapidly with a higher r.

3. (a) Answers may vary. Increasing the initial value size increases the output values.

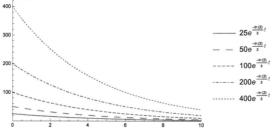

(b) Yes. Answers may vary. Changing the initial value size increases the rate of change of the amount.

(c) $A(t)$ has been multiplied by a factor greater than 1 if the initial size is larger. $A(t)$ has been multiplied by a positive factor less than 1 if the initial sample size is smaller.

4. (a) $h'(t) = -19.6t + v_0$. $h'(0) = -19.6(0) + v_0 = v_0$.

(b) Answers will vary. The initial velocity affects the maximum height of the ball. See figure below.

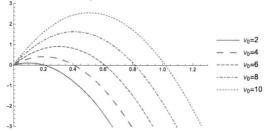

(c) The ball is in the air for $\frac{v_0}{9.8}$ seconds.

5. (a) See figure below.

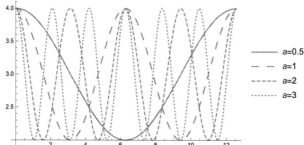

(b) a affects the period of $h(t)$.

(c) a is dependent on how fast the Penguin spins his umbrella.

6. (a) Answers may vary but should discuss how the location of the inflection point is moved to the left or right. See figure below.

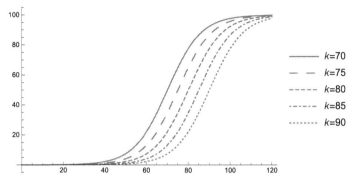

(b) Answers may vary but should discuss how the end behavior (and horizontal asymptote) of the function changes with changes in C. See figure below.

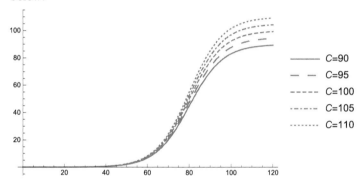

7. (a) Answers may vary but should discuss how the location of the first peak corresponds to the value of c. See figure below.

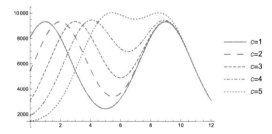

(b) Answers may vary but should discuss how the steepness of the curve changes overall and how the curve changes near the relative minimum.

See figure below.

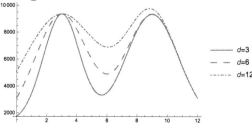

8. (a) 3.474% is in the system.

    (b) 19.285% is in the system.

    (c) 2.901% is in the system.

    (d) After 12 hours, 3.474% of acetaminophen ($k = 0.28$) is left in the system and 2.901% of ibuprofen ($k = 0.295$) is left in the system. With a change in k of 0.015 the resulting percentage change is 0.572%. One can also see that if $k = 0.285$ that the percentage change would be 0.202%. This result is not particularly sensitive to a small change in k.

2.1 Solutions

1. Answers will vary.

2. (a) $P'(t) = x^3 - 16x^2 + 60x$

    (b) $P'(10) = 0$, which does not tell us if profits will grow after the end of year 10. This number alone does not tell us what is happening for $t > 10$. Finding $P'(10.1)$ or $P'(10.2)$ may help provide more information.

    (c) $P'(11) = 55$, which means that at the end of year 11, the model indicates that profits will be increasing at a rate of 55 thousand dollars per year. If the model is a good fit to a significant amount of data, the company probably should continue to make toys.

3. (a) See figure below.

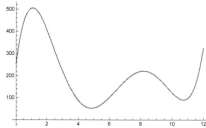

(b) See figure below.

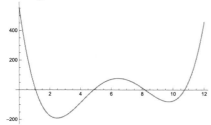

(c) The balance is increasing on about $(0,1.1) \cup (4.8,8.2) \cup (10.7,12)$. In other words, his balance is increasing at the beginning of the year until about the beginning of February, from about the end of May to a little into September, and about the middle of November to the end of the year. The balance is decreasing on about $(11.1,4.8) \cup (8.2,10.7)$. In other words, his balance is decreasing from about the beginning of February until about the end of May and from a little into September until midway through November.

(d) The balance is increasing and the rate of increase is increasing on about $(4.8,6.5) \cup (10.7,12)$. These intervals correspond to near the end of May until halfway into July and from late November until the end of the year.

4. (a) $A(t)$ is always concave down, which means that $A'(t)$ is decreasing on the entire domain (of more than 12 years) shown.

(b) The fact that $A'(t)$ is decreasing means that the rate of change is getting smaller with time. When the slope of the tangent line is at its steepest, the rate of decrease is the greatest, which is probably around the final payment (about 12.5 years after the money was borrowed).

5. (a) 8.5 mph

(b) (2,4) and (4,10)

(c) At 11 minutes.

2.2 Solutions

1. 4.39 feet by 4.39 feet.

2. (a) $x \approx 1.07$. His balance is highest at in early February.

(b) $x \approx 4.85$. His balance is the lowest near the end of May.

(c) Although the derivative has a local maximum around mid-July, $A'(0)$ and $A'(12)$ are both higher than the derivative output here. Todd's credit card balance is increasing most rapidly at the beginning of the year (assuming that the curve is representative at an endpoint).

   (d) $x \approx 2.47$. His balance is decreasing most rapidly about halfway into March.

3. (a) The points of John's clavicle that the doctor is examining are $(\frac{3}{2},0)$ and $(\frac{5}{2},0)$.

   (b) The top of John's head is at the point $(2,8)$.

   (c) The points of inflection (and injection) will be at roughly $(1.77639,4.096)$ and $(2.22361,4.096)$.

4. (a) The fox population is rapidly increasing because of the large food source population.

   (b) The chicken population is rapidly decreasing because there are a lot of foxes around to eat them.

   (c) At $t = 2.362671$, the chicken population is decreasing at its fastest rate. That is, the rate of change of the chicken population is minimized when $t = 2.362671$.

   (d) Each time the chicken (or fox) population reaches a size of 341, the population is decreasing or increasing at its fastest rate.

5. (a) $y \approx 0.636$

   (b) George will be working his hardest when $x = 0.691$ and $y = 1.3827$.

2.3 Solutions

1. The probability of fatality between 4 and 9 km from the earthquake's epicenter is 0.415.

2. (a) $A'(t) = 5e^{0.01t}$

   (b) 5.03. This number represents the amount of interest in dollars accumulated in the first year.

3. (a) The probability that a random woman's pregnancy would last less than 268 days is 50%.

   (b) The chance that a random woman's pregnancy would last between 270 and 280 days is approximately 24%.

4. About 15 mg.

2.4 Solutions

1. (a) $\pi \left(45 + 2 \int_3^6 x \left(\frac{10 + 3x}{\sqrt{x-2}} - 14 \right) dx \right)$

   (b) $\frac{1121\pi}{15}$ in^3.

2. 7.41 in^3.

3. The volume of Ben's new design is $\frac{27\pi}{14}$ in^3 which is approximately 6.059 in^3 and close to the volume (approximately 6.103 in^3) of the shop's current design.

4. (a) The volume of the clarinet is 3.292 in^3.

   (b) There are many solutions. However, one example that satisfies the trumpet bugles' dimensions is the function $f(x) = \frac{13}{4\sqrt{x}} - \frac{1}{4}$. The volume of this trumpet bugle is 78.478 in^3.

   (c) Again there are many solutions. One sample that would satisfy the tuba bugle's dimensions is $f(x) = \frac{45}{4\sqrt[4]{x}} - \frac{11}{4}$. The volume of this tuba bugle is 1656.27 in^3.

   (d) Many solutions exists. For example, Gabriel's horn formed by taking $\frac{1}{x}$ on $[1, \infty)$ forms a horn with volume of π and an infinite surface area.

5. (a) The volume of the first flotation device is approximately 0.142 ft^3.

   (b) The volume of the second flotation device is approximately 20.1062 ft^3.

   (c) The volume of the third flotation device is $\frac{\pi}{2}$ ft^3.

   (d) The buoyancy in newtons for each device is 39.52, 174.54, and 436.35, respectively.

   (e) In pounds of force the devices provide 8.88, 39.24, and 98.09 buoyancy, respectively. Thus, devices 2 and 3 would be safe for use on children based on the regulations.

2.5 Solutions

1. (a) The sequence is divergent.

   (b) The sequence is divergent.

   (c) As $n \to \infty$ the points are approaching being on an ellipse.

2. (a) The sequence is converging to $\frac{3}{2}$.

(b) The sequence is divergent.

(c) The points are approaching being on a line segment where $x = \frac{3}{2}$ and y is between -2 and 2.

3. (a) $p_{10} \approx 46.5$.

   (b) The pangolin population will converge.

   (c) As $n \to \infty$, the pangolin population converges to 80 pangolin.

4. (a) $s_1 \approx 1.9$ and $f_1 \approx 11.05$.

   (b) Neither the predator nor the prey populations will converge.

5. (a) $p_{10} \approx 29$ pangolin.

   (b) The population of pangolin diverges.

6. (a) Russell should choose the ticket that pays $800 initially and for 7 subsequent years.

   (b) Russell would need to live the 7 subsequent years and be paid in full for the winnings in order to make more money with his choice.

7. (a) $1 - \frac{364}{365} \approx 0.00274$

   (b) $\frac{365}{365} \frac{364}{365} \frac{363}{365} \approx 0.9918$

   (c) The probability that no two people in a group of size $n < 365$ share the same birthday is $\displaystyle\prod_{i=0}^{n} \frac{365 - i}{365}$.

   (d) If Sam and 22 others are in the room together then the probability that at least two people share the same birthday is approximately 0.5073 and thus there is a more than 50% chance that this will happen.

   (e) This sequence of probabilities converges to 1.

8. (a) The square has side length of approximately 2.93004.

   (b) The first three positions of the red point are $(\frac{2.93004}{2}, 1.93004)$, $(2.93004, 0.465019)$, and $(\frac{3 \cdot 2.93004}{2}, -1.77378)$.

   (c) The wheel will roll in a repetitive nature and thus the sequence of the positions of the red point will never converge.

3.1 Solutions

1. (a) Equity Autos should purchase 10 comedy commercials and 5 football commercials.

(b) If Equity Autos wants to reach c million high-income women and c million high-income men, they should purchase $c/8$ comedy commercials and $c/16$ football commercials. For the answer to be physically reasonable, c must be divisible by 16.

2. (a) The solution is $x = \begin{pmatrix} 1 \\ 1 \\ 1 \end{pmatrix}$.

(b) The solution to part (a) tells us that Juicemathics needs 18 ounces of cranberry juice, 22 ounces of orange juice, and 8 ounces of pineapple juice in order to make one 16 ounce bottle of each of the mixed juices.

3. (a) JBJ would need 11 square feet of denim, 7 packs of brads, and 23 spools of thread to make 1 pair of jeans, 2 pairs of overalls, and 3 hats.

(b) JBJ could make 9 hats if they purchase 69 spools of thread and adequate amounts of denim and brads.

(c) Yes, JBJ could make exactly 159 jeans, 32 pairs of overalls, and 11 embroidered hats.

4. (a) $A = \begin{pmatrix} 500 & 200 \\ 150 & 500 \end{pmatrix}, b = \begin{pmatrix} 9000000 \\ 6000000 \end{pmatrix}$

(b) The populations that produce the economies in part (a) are Exponentitown 15,000 citizens and Logicity 7,500 citizens.

(c) The nullspace of A^T is orthogonal to the rowspace of A. $(21000, x)$ should be in the rowspace of A where x is Logicity's income. A basis vector for the nullspace of A^T will be orthogonal to $(21000, x)$ and thus Logicity's income will be $10,500.

5. (a) One solution is that $x_1 = 1, x_2 = 5, x_3 = 4$, and $x_4 = 3$. Any linear combination of solutions is again a solution.

(b) 0.4 moles C_3H_8 and 2 moles O_2.

6. (a) Solving the system presented here is the same as finding the eigenvector, of the coefficient matrix, associated with the eigenvalue of magnitude 1. This eigenvector presented as a percentage vector is approximately $\{.28, .35, .37\}$ and thus approximately 28% of the economy must be made up of p_1 in order to sustain this closed economy.

(b) Approximately 80 units of p_1 and 106.67 units of p_3.

7. China should produce approximately 35.15% of the commodities, India should produce 49.03% and Singapore should produce approximately 15.81%.

3.2 Solutions

1. The $(8,8)^{\text{th}}$ entry equal to 1 represents an absorbing state and thus eventually the entire population will end up in the 8th state.

2. (a) This is an absorbing state which means that once a student graduates they stay in this state.

   (b) After 4 years, $T^4 \cdot \begin{pmatrix} 0 \\ 100 \\ 200 \\ 0 \end{pmatrix} = \begin{pmatrix} 61.75 \\ 0.81 \\ 3.57 \\ 233.87 \end{pmatrix}$ students are in each state.

   So approximately 233 students have graduated after 4 years.

   (c) The $(2,2)$ entry should be 1.3.

3. (a) The transition matrix is $\begin{pmatrix} 0.9 & 0.25 & 0.6 \\ 0.1 & 0.45 & 0 \\ 0 & 0.3 & 0.4 \end{pmatrix}$ where the first column represents the movement from the "not-enrolled" state, the second column represents movement from the "white belt" state, and the third column represents movement from the "yellow belt" state.

   (b) At the beginning of the program, 30 students are enrolled in white belt status. At the beginning of the second month, there will be approximately 41 students enrolled in white belt status and 9 enrolled in yellow belt status.

4. (a) Approximately 1,250 players will be on level 3 after 5 plays.

   (b) If we start with 10,000 players at the bottom, in the long run approximately 5,000 will be at the bottom, 2,500 at level 2, 1,250 at level 3, and 625 at levels 4 and 5 each.

5. The transition matrix is $\begin{pmatrix} 0 & \frac{3}{10} & 0 & 0 \\ 1 & \frac{3}{10} & \frac{5}{11} & 0 \\ 0 & \frac{4}{10} & \frac{5}{11} & \frac{1}{2} \\ 0 & 0 & \frac{1}{11} & \frac{1}{2} \end{pmatrix}$ and in the long run approximately 12.82% of pitches will be 60, 42.73% will be 62, 37.61% will be 64, and 6.84% will be 70.

6. Approximately 37.35% of freshman are single after 2 months. At the end of freshman year (after 9 months), approximately 36.42% of freshman are single. Although there is a slight difference in the percentage of single freshman after 9 months if the initial number of single freshman changes, it is not significant.

7. After 5 years there are 576 age 0 birds, 151 age 1 birds, and 60 age 2 birds. After 100 years the bird population has exploded to approximately 727,318 age 0 birds, 135,197 age 1 birds, and 62,827 age 2 birds.

8. The long run behavior is fairly sensitive to the initial conditions. If we start with 99 birds in each of the 3 categories we end up with approximately 720,045 age 0, 133,845 age 1, and 62,199 age 2 birds and if we start out with 101 birds in each category we end up with approximately 734,591 age 0, 136,548 age 1, and 63,455 age 2 birds. If there are only 90 birds initially in each category we end up with 654,586 age 0, 121,677 age 1, and 56,544 age 2 birds and with 110 birds initially in each category 800,050 age 0, 148,716 age 1, and 69,110 age 2 birds in the long run.

3.3 Solutions

1. The feasibility region can be see in the figure below. The corners of the feasibility region are $(1,0),(2.5,0)$, and $(\frac{13}{7},\frac{6}{7})$. The objective function $2x_1 + x_2$ has values of $2, 5$, and $\frac{32}{7}$, respectively, at each of these corners and thus reaches it's maximum when $x_1 = 2.5$ and $x_2 = 0$.

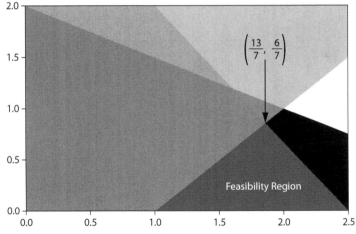

2.

$$\begin{pmatrix} x_1 & x_2 & x_3 & s_1 & s_2 & s_3 & C & constants \\ 3 & 1 & 1 & 1 & 0 & 0 & 0 & 60 \\ 1 & -1 & 2 & 0 & 1 & 0 & 0 & 10 \\ 1 & 1 & -1 & 0 & 0 & 1 & 0 & 20 \\ \hline 2 & -1 & 1 & 0 & 0 & 0 & 0 & 0 \end{pmatrix}$$

where $s_1, s_2, s_3 \geq 0$.

3. If x_1 represents the number of robotic claws made and x_2 represents the number of glo specs, the constraints mentioned are $1 \leq x_1, 1 \leq x_2, x_1 + x_2 \leq 20, x_2 \leq x_1$. The feasibility region can be seen in the figure below.

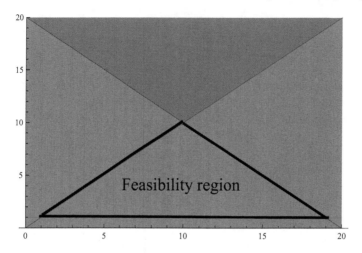

4. If c is the number of calculus classes and s is the number of statistics classes offered, then the tableau is

$$\begin{pmatrix}
c & s & s_1 & s_2 & C & constants \\
\hline
1 & 1 & 1 & 0 & 0 & 15 \\
35 & 30 & 0 & 1 & 0 & 480 \\
\hline
-1 & -1 & 0 & 0 & 1 & 0
\end{pmatrix}$$

where $s_1 \geq 0$ and $s_2 \leq 0$. Since the negative numbers in the last row are of the same size, either column 1 or column 2 can be chosen as the pivot column. We will use column 1 and thus the first pivot element is the $(2,1)$ entry. Using row operation to make that entry a 1 we get

$$\begin{pmatrix}
c & s & s_1 & s_2 & C & constants \\
\hline
0 & \frac{1}{7} & 1 & -\frac{1}{35} & 0 & \frac{9}{7} \\
1 & \frac{30}{35} & 0 & \frac{1}{35} & 0 & \frac{480}{35} \\
\hline
0 & -\frac{1}{7} & 0 & \frac{1}{35} & 1 & \frac{480}{35}
\end{pmatrix}.$$

The next pivot will come from column 2, with a final tableau of

$$\begin{pmatrix}
c & s & s_1 & s_2 & C & constants \\
\hline
0 & 1 & 7 & -\frac{7}{35} & 0 & 9 \\
1 & 0 & -6 & \frac{1}{5} & 0 & 6 \\
\hline
0 & 0 & 1 & 0 & 1 & 15
\end{pmatrix}$$

with $s_1 = s_2 = 0$, $c = 9$, $s = 6$ and C=number of classes=15.

5. If c is the units of cereal and m is the units of milk, the constraints in this problem are

$$\begin{aligned}
3c + m &\geq 8, \\
4c + 3m &\geq 19, \\
c + 3m &\geq 7, \\
c &\geq 0, m \geq 0,
\end{aligned}$$

while the objective function is $2c + 4m$. In order to meet all of the dietary needs while keeping your cost at a minimum less than or equal to $20, 4 cereal units and 1 milk unit should be purchased.

6. If h is the number of halloween packs and k is the number of knight packs, the constraints for the toy brick company are

$$
\begin{aligned}
12k + 18h &\leq 100 \\
3k + h &\leq 20 \\
k \geq 0, \quad h &\geq 0,
\end{aligned}
$$

with an objective function of $6h + 4k$ to maximize. The toy brick company should make 5 knight packs and 2 halloween packs for a revenue of $32.

7. If h is the number of halloween packs and k is the number of knight packs, the constraints for the toy brick company are

$$
\begin{aligned}
12k + 18h &\leq 100 \\
3k + 2h &\leq 20 \\
4k + 5h &\leq 25 \\
k \geq 0, \quad h &\geq 0,
\end{aligned}
$$

with an objective function of $6h + 4k$ to maximize. The toy brick company should make 0 knight packages and 5 halloween packs for a revenue of $30.

3.4 Solutions

1. A line will not be the best curve to model this data as a line with positive slope grows infinitely large and in this problem we are trying to predict probability (so the dependent variable should remain less than or equal to 1).

2. (a) 57.2% support for Trump in Michigan

   (b) 5.2% support for Trump in DC

   (c) A linear model can take on negative values or values greater than 100%. In this problem, we really want a model where the dependent variable only takes on values between 0 and 100%.

3. (a) $\widehat{graduates} = 56098.1 + 0.000974099(funds)$

   (b) The local agencies should be funded approximately $712,000,000.

4. (a) If the years are shifted, making year 1960 (t=0) then the least squares linear regression model is $\widehat{expectancy} = 65.646 + 0.208t$.

   (b) Using the regression line, Tim's life expectancy ($t = 13$) is predicted to be approximately 68.355 years or 68 years and a little over 4 months.

5. The least squares third degree polynomial fit for the Zika data can be found using the matrix

$$A = \begin{pmatrix} (6{,}828{,}065)^3 & (6{,}828{,}065)^2 & 6{,}828{,}065 & 1 \\ (2{,}911{,}641)^3 & (2{,}911{,}641)^2 & 2{,}911{,}641 & 1 \\ (4{,}670{,}724)^3 & (4{,}670{,}724)^2 & 4{,}670{,}724 & 1 \\ (9{,}922{,}576)^3 & (9{,}922{,}576)^2 & 9{,}922{,}576 & 1 \\ (2{,}992{,}333)^3 & (2{,}992{,}333)^2 & 2{,}992{,}333 & 1 \\ (10{,}042{,}802)^3 & (10{,}042{,}802)^2 & 10{,}042{,}802 & 1 \\ (11{,}613{,}423)^3 & (11{,}613{,}423)^2 & 11{,}613{,}423 & 1 \\ (4{,}896{,}146)^3 & (4{,}896{,}146)^2 & 4{,}896{,}146 & 1 \\ (2{,}995{,}919)^3 & (2{,}995{,}919)^2 & 2{,}995{,}919 & 1 \\ (626{,}042)^3 & (626{,}042)^2 & 626{,}042 & 1 \end{pmatrix}$$

and gives the model $\hat{y} = -1.1643 + 3.17893 * 10^{-6}x - 1.40907 * 10^{-13}x^2 + 1.40737 * 10^{-21}x^3$. For this cubic model the maximum error is 5.0338, the l_2 error is 6.99, and the relative l_2 error is 18%. Based on these numbers, this model may be a slightly better fit than the quadratic model and the linear model.

6. (a) $A = \begin{pmatrix} 6828065 & 1 \\ 2911641 & 1 \\ 4670724 & 1 \\ 9922576 & 1 \\ 2992333 & 1 \\ 10042802 & 1 \\ 11613423 & 1 \\ 4896146 & 1 \\ 2995919 & 1 \\ 626042 & 1 \end{pmatrix}$

   (b) $\hat{y} = 3.46573e^{(1.58239*10^{-7})x}$

3.5 Solutions

1. (a) The dissimilarity cost is 14.

   (b) $D = \begin{pmatrix} 3 & 5 & 4 \\ 5 & 4 & 5 \\ 4 & 5 & 1 \end{pmatrix}$ with dissimilarity cost 14.

2. b and c

3. You are given a problem with 14 records to analyze and you wish to test each of the permutations with a cost function, for example, embedded zeros or dissimilarity. The computation you will have to do over 14! computations, or over 87,178,291,200 computations, which is computational time cumbersome.

4. (a) (1234),(1243),(1342),(1324),(1423),(1432),(1)(234),(1)(243),
    (2)(134),(2)(143),(3)(124),(3)(142),(4)(123),(4)(132),(12)(34),
    (12)(3)(4),(13)(24),(13)(2)(4),(14)(23),(14)(2)(3),(1)(4)(23),
    (1)(3)(24),(1)(2)(34),(1)(2)(3)(4) with a total of 24 unique permutations.

   (b) 6,5,5,4,1,1,2,1,6,2,5,1,6,5,4,4,4,6,1,2,2,1,1,1, respectively.

   (c) As one example, permutation (1423) gives a minimum number of embedded zeros, 1. The ordering in this case is Wanna Dance, All of You, Billie Jean, Like a Virgin.

5. $S = \begin{pmatrix} 0 & 0 & 0 & 0 \\ 0 & 3 & 2 & 2 \\ 0 & 2 & 4 & 3 \\ 0 & 2 & 3 & 4 \end{pmatrix}$ and $D = \begin{pmatrix} 5 & 5 & 5 & 5 \\ 5 & 2 & 3 & 3 \\ 5 & 3 & 1 & 2 \\ 5 & 3 & 2 & 1 \end{pmatrix}$.

6. For the permutation $(1)(2)(3)(4)$, S and D can be found in the solution 2.5 #5 and the dissimilarity cost is 15. For permutation $(1)(234)$,

   $S = \begin{pmatrix} 0 & 0 & 0 & 0 \\ 0 & 4 & 2 & 3 \\ 0 & 2 & 3 & 2 \\ 0 & 3 & 2 & 4 \end{pmatrix}, D = \begin{pmatrix} 0 & 0 & 0 & 0 \\ 0 & 4 & 2 & 3 \\ 0 & 2 & 3 & 2 \\ 0 & 3 & 2 & 4 \end{pmatrix}$ and the dissimilarity cost

   is 16. Based just on these two permutations we would say the permutation $(1)(2)(3)(4)$ gives the minimum dissimilarity cost and the song ordering should be Like a Virgin, Billie Jean, Wanna Dance, All of You.

3.6 Solutions

1. $U = \begin{pmatrix} \frac{1}{\sqrt{2}} & 0 & -\frac{1}{\sqrt{2}} \\ \frac{1}{\sqrt{2}} & 0 & \frac{1}{\sqrt{2}} \\ 0 & 1 & 0 \end{pmatrix}$, $W = \begin{pmatrix} 2 & 0 & 0 \\ 0 & 1 & 0 \\ 0 & 0 & 0 \end{pmatrix}$, $V = \begin{pmatrix} \frac{1}{\sqrt{2}} & 0 & -\frac{1}{\sqrt{2}} \\ \frac{1}{\sqrt{2}} & 0 & \frac{1}{\sqrt{2}} \\ 0 & 1 & 0 \end{pmatrix}$

2. $U = \begin{pmatrix} \frac{1}{\sqrt{2}} & 0 \\ \frac{1}{\sqrt{2}} & 0 \\ 0 & 1 \end{pmatrix}$, $W = \begin{pmatrix} 2 & 0 \\ 0 & 1 \end{pmatrix}$, $V = \begin{pmatrix} \frac{1}{\sqrt{2}} & 0 \\ \frac{1}{\sqrt{2}} & 0 \\ 0 & 1 \end{pmatrix}$

3. 2.91, 1.24, 1, 0(multiplicity 2)

4. (a) The advertisement-appeals matrix is $A = \begin{pmatrix} 0 & 1 & 0 & 1 & 1 & 0 \\ 0 & 1 & 1 & 0 & 0 & 0 \\ 0 & 1 & 0 & 1 & 0 & 0 \\ 0 & 1 & 0 & 1 & 1 & 1 \end{pmatrix}.$

Using singular value decomposition, $A = UWV^T$ where

$$U = \begin{pmatrix} 0.572583 & 0.136444 & -0.257443 & -0.766325 \\ 0.259709 & -0.913541 & 0.304875 & -0.071028 \\ 0.42611 & -0.158861 & -0.715422 & 0.530438 \\ 0.650485 & 0.348696 & 0.573537 & 0.355438 \end{pmatrix},$$

$$W = \begin{pmatrix} 2.88966 & 0. & 0. & 0. & 0. & 0. \\ 0. & 1.28173 & 0. & 0. & 0. & 0. \\ 0. & 0. & 0.830779 & 0. & 0. & 0. \\ 0. & 0. & 0. & 0.5629 & 0. & 0. \end{pmatrix},$$

and

$$V^T = \begin{pmatrix} 0. & 0.661 & 0.090 & 0.571 & 0.423 & 0.225 \\ 0. & -0.458 & -0.713 & 0.255 & 0.379 & 0.272 \\ 0. & -0.114 & 0.367 & -0.481 & 0.380 & 0.690 \\ 0. & 0.086 & -0.126 & 0.212 & -0.730 & 0.631 \\ 0. & -0.577 & 0.577 & 0.577 & 0. & 0. \\ 1. & 0. & 0. & 0. & 0. & 0. \end{pmatrix}.$$

(b) $\hat{A} = \hat{U}\hat{W}\hat{V}^T$ where $\hat{U} = \begin{pmatrix} 0.572583 & 0.136444 \\ 0.259709 & -0.913541 \\ 0.42611 & -0.158861 \\ 0.650485 & 0.348696 \end{pmatrix},$

$$\hat{W} = \begin{pmatrix} 2.88966 & 0. \\ 0. & 1.28173 \end{pmatrix},$$

and $\hat{V}^T = \begin{pmatrix} 0. & 0.661 & 0.090 & 0.571 & 0.423 & 0.225 \\ 0. & -0.458 & -0.713 & 0.255 & 0.379 & 0.272 \end{pmatrix}.$

(c) In the figure below, we can see the singular decomposition using 2 singular values. Here you see that Argentina and USA are most similar.

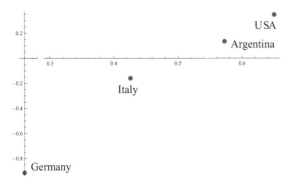

4.2 Solutions

1. (a) $\dfrac{dA}{dt} = 165 + 0.045 * A$

   (b) $A(t) = 7666.67e^{0.045t} - 3666.67$

   (c) See figure below.

   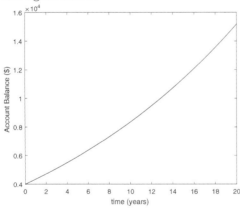

   (d) The account will have \$12,000 after approximately 190.68 months (15.89 years).

2. (a) *Answers may vary* $\dfrac{dS}{dt} = \dfrac{3}{5} - 0.45 * S; S(0) = 4$

   (b) $S(t) = 1.33 + 2.67e^{-0.45t}$

(c) See figure below.

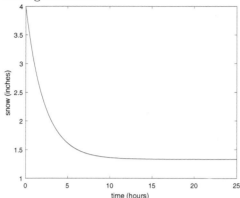

(d) Under the presented assumptions, it is not possible for the snow to be cleared while it is still snowing. The model predicts that the fleet can maintain a level of 1.5 inches while it is still snowing.

3. (a) *Answers may vary* $\dfrac{dX}{dt} = D - kX$, where D represents the hourly dosage, and k is the clearance proportionality constant.

(b) $X(t) = 71.4286 - 71.4286e^{-0.35t}$ (depicted in figure below)

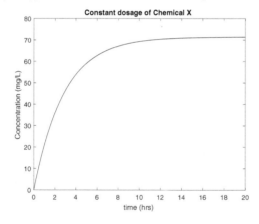

(c) $X(t) = 400e^{-0.35t}$ (depicted in figure below)

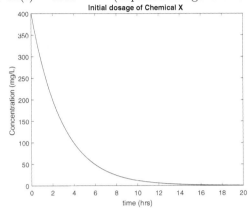

Initial dosage of Chemical X

(d) In order to ensure that the patient has at least 60 mg/L after 10 hours, the doctor should employ the constant dosing strategy.

4. (a) *Answers may vary* $i(t) = \dfrac{9}{5} - \dfrac{9}{5}e^{-20t}$

   (b) See figure below.

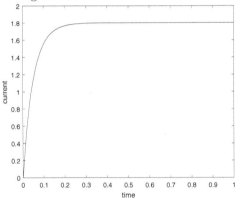

   (c) As time increases, the current across the circuit goes to $9/5$.

5. (a) *Answers may vary* $\dfrac{dT}{dt} = \lambda + \alpha T - \beta T^2 - \delta; T(0) = 2$

   (b) Applying the given parameter values gives the initial value problem, $\dfrac{dT}{dt} = 0.015 + 2T - .75T^2; T(0) = 2$, whose solution is depicted in part 5c.

(c) See figure below.

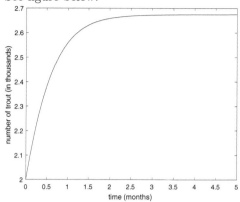

(d) The fish population increases to approximately 2,675 where it appears to reach equilibrium. To change the outcome of the population (make it decrease), we would want to increase the value of β.

6. (a) See figure below.

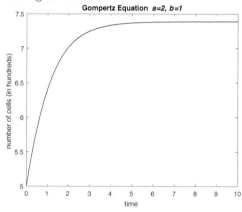

(b) See figure below.

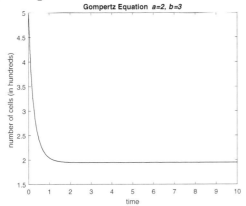

(c) See figure below.

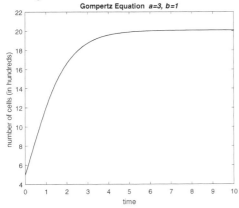

(d) From these few simulations, it appears that the population grows when $a > b$, and decreases when $a < b$. Greater differences between the parameter values result in larger increases or decreases in the population.

4.3 Solutions

1. (a) As shown in the figure below, there appears to be an initial competition between the two sodas, where they both have a decrease in sales, but after 10 days, Calcu-Cola is selling around 5000 while Pepsilon has

been driven out of the market.

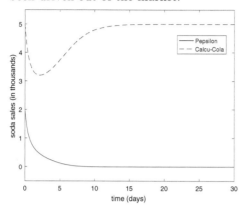

(b) As shown in the figure below, there appears to be an initial competition between the two sodas, where they both have a decrease in sales, but after 10 days, Pepsilon is selling around 3200 while Calcu-Cola has been driven out of the market.

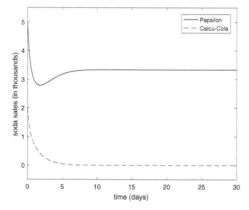

(c) As shown in the figure below, there appears to be an initial competition between the two sodas, where they both have a decrease in sales, but after 10 days, Pepsilon is selling around 3200 while Calcu-Cola has been driven out of the market.

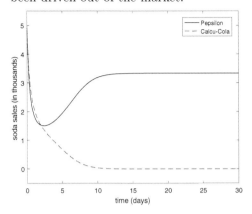

2. (a) $S(t) = 25 + 25e^{-1.35t}$, $\quad C(t) = 25 - 25e^{-1.35t}$

(b) The concentrations will never be the exact same; however, their difference is less than 10^{-6} after 13.14 seconds. Note that both concentrations approach 25 μM as $t \to \infty$.

(c) See figure below.

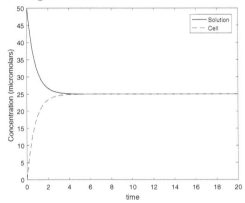

3. (a) The results from the first model, with $\alpha = 1 \times 10^{-6}$ and $\beta = 1 \times 10^{-7}$, are depicted in the figure below. These results suggest that virtually all of the Union Army was killed, and so the Confederate Army would have won the battle. This didn't match with the historical facts, so

Sean knew that there was an error somewhere.

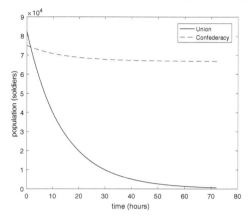

(b) With the new set of parameters, the results (see figure below) seem more realistic. The Union Army maintains an advantage over the Confederates throughout the battle. At the conclusion of the battle, the model predicts 23,474 casualties for the Union Army and 29,343 casualties for the Confederate Army. Both predictions are similar to the historical data.

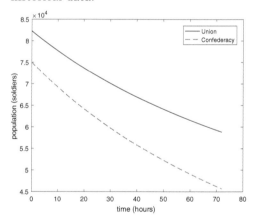

4. (a)

$$\frac{dR}{dt} = aR - bRW$$

$$\frac{dW}{dt} = -dW + cRW$$

(b) *Answers may vary* aR represents the rate at which rabbits enter the park, either through birth or migration. bRW is the rate at which rabbits leave the park, either through death or migration. cRW is the rate at which wolves enter the park, most likely through migration.

dW is the rate at which wolves leave the park, either through death or migration.

(c) The solution to the system of differential equations is shown in the figure below. Notice that as the wolf population increases, the population of rabbits dramatically decreases. Then, when there are little to no rabbits present, the population of wolves subsides. When the wolf population has significantly decreased, the rabbits come back. The cycle continues throughout the 3 year period.

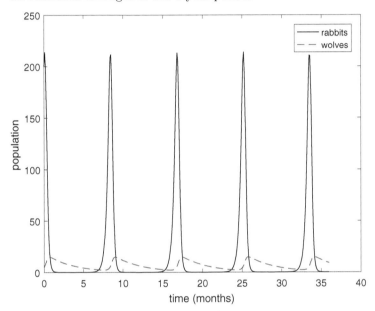

5. (a) See figure below.

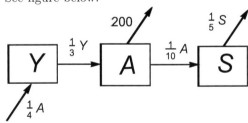

(b)

$$\frac{dY}{dt} = \frac{1}{4}A - \frac{1}{3}Y$$

$$\frac{dA}{dt} = \frac{1}{3}Y - \frac{1}{10}A - 200$$

$$\frac{dS}{dt} = \frac{1}{10}A - \frac{1}{5}S$$

(c) The figure below displays the solution to the system with 200 deer being killed per year. Examination of the figure shows that the populations of all three groups reach zero by year 15. The population is being overhunted, which will lead to the population dying out. The reserve will not be able to stay in business under these conditions.

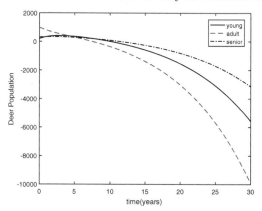

(d) The figure below displays the solution to the system when the number of deer hunted is reduced to 100. Under this assumption, there is an initial decrease in the adult population, but it rebounds, and all three groups of deer have increasing populations after 30 years. This seems to ensure that the reserve can stay in business long term.

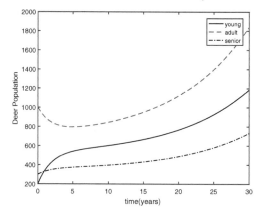

6. (a) See figure below.

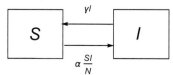

(b)

$$\frac{dS}{dt} = \gamma I - \alpha \frac{SI}{N}$$

$$\frac{dI}{dt} = \alpha \frac{SI}{N} - \gamma I$$

(c) See figure below.

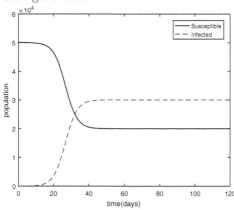

4.4 Solutions

1. (a) $x(t) = \dfrac{1}{2}\cos 4t - \dfrac{1}{4}\sin 4t$

(b) See figure below.

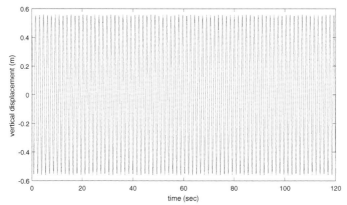

The toy continues to oscillate for the entire time.

2. (a) $x(t) = \dfrac{1}{2}e^{-2t}(\cos 3.4641t)$

(b) See figure below.

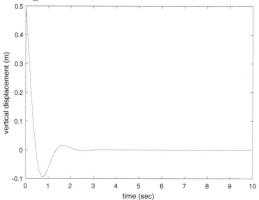

Underwater, there are minimal oscillations. The toy returns to rest after approximately 3.5 seconds.

3. (a) See figure below.

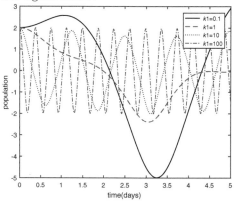

(b) As $k1$ increases, the amplitude of the oscillations decreases, and the frequency increases.

4. (a) See figure below.

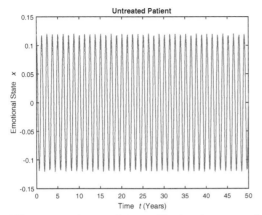

The patient's mood is oscillating between -0.12 and 0.12, indicating a shift from a depressive state to a hypomanic state. The oscillations are consistent and have the same amplitude over the 50 year period.

(b) See figure below.

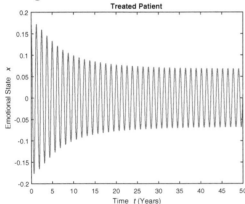

The patient's mood starts with oscillations between -0.18 and 0.18, but the severity of the oscillations diminish, and from about year 15 forward, the amplitude of the oscillations is about 0.07. This implies that the patient still has mood swings, but the severity is lessened while treated.

5. (a) See figure below

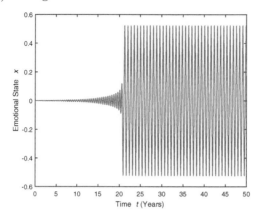

(b) The patient's mood starts with large oscillations between −0.2 and 0.2, but the severity of the oscillations diminishes, and from about year 40 forward, the amplitude of the oscillations is about 0.03. This implies that the patient still has mood swings, but the severity is lessened with treatment, likely in the normal range.

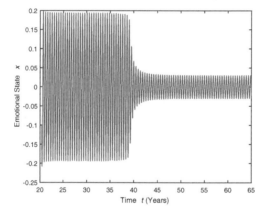

5.1 Solutions

1. See figure below.

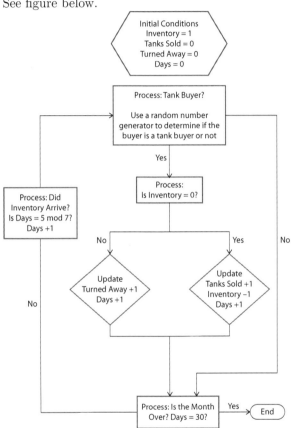

2. Technology dependent.

3. There is a huge assumption in the flowchart presented here. That is that all elephants that you hit are not already sterilized. Some students will also include this issue in their flowchart and their simulation in the next exercise.

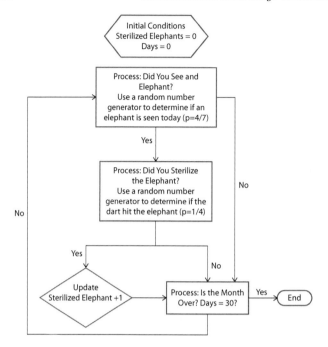

4. Technology dependent.

5. Jack's expected winnings are $10\left(\frac{4}{6}\right)^{10} + 8\left(\frac{2}{6}\right)\left(\frac{4}{6}\right)^9 + 6\left(\frac{2}{6}\right)^2\left(\frac{4}{6}\right)^8 + 4\left(\frac{2}{6}\right)^3\left(\frac{4}{6}\right)^7 + 2\left(\frac{2}{6}\right)^4\left(\frac{4}{6}\right)^6 + 0\left(\frac{2}{6}\right)^5\left(\frac{4}{6}\right)^5 - 2\left(\frac{2}{6}\right)^6\left(\frac{4}{6}\right)^4 - 4\left(\frac{2}{6}\right)^7\left(\frac{4}{6}\right)^3 - 6\left(\frac{2}{6}\right)^8\left(\frac{4}{6}\right)^2 - 8\left(\frac{2}{6}\right)^9\left(\frac{4}{6}\right) - 10\left(\frac{2}{6}\right)^{10} \approx .28$.

6. If the probability of death decreases from 80% then the number of Wolbachia females will grow more slowly and the number of susceptible female mosquitoes will take longer to get to 0.

7. Technology dependent.

5.2 Solutions

1. (a) $p_0 = 10, p_1 = 15, p_2 = 20, p_3 = 25, p_4 = 30, p_5 = 35, p_6 = 40, p_7 = 47.5, p_8 = 57.5, p_9 = 70, p_{10} = 85, p_{11} = 102.5, p_{12} = 122.5, p_{13} = 146.25, p_{14} = 175, p_{15} = 210$.

   (b) The population grows by 20% percent between $t = 14$ and $t = 15$.

   (c) If the time step is 140 days, then $p_0 = 10$ and $p_n = 1.2p_{n-1}$.

   (d) 252 female pangolins.

2. Technology dependent.

3. Again, assuming that all elephants hit are not previously sterilized, $e_{n+1} = e_n + 1$.

4. Technology dependent.

5. $Y0_{n+1} = 4Y1_n + 6Y2_n$, $Y1_{n+1} = .2Y0_n$, $Y2_{n+1} = .5Y1_n$.

6. The initial effect of increasing parameter a is that the shark population will die out faster and the fish population will grow faster. However, a small shark population may then thrive in this situation and we may still see an oscillating behavior (depending on how large a is).

7. The Process: Will Shark Eat? is directly related to the Process: Will Fish Die? in the fish model behavior.

8. Technology dependent.

5.3 Solutions

1. (a) See table below.

		Friend	
		Comedy	Action Thriller
You	Comedy	(3,2)	(1,1)
	Action Thrillers	(0,0)	(2,3)

(b) The mixed-strategy equilibrium for this game is for you to pick the comedy with probability $\frac{3}{4}$ and the action thriller with probability $\frac{1}{4}$, while your friend picks the comedy with probability $\frac{1}{4}$ and the action thriller with probability $\frac{3}{4}$.

2. The mixed strategy for the kicker is that they will choose to kick left with probability $\frac{1}{2}$ and choose to kick right with probability $\frac{1}{2}$. The goalkeeper will also have a mixed strategy of diving right with probability $\frac{1}{2}$ and left with the same probability. The kicker's expected utility is $\frac{1}{2}$ and the goalkeeper's expected utility is $\frac{1}{2}$ and thus we have a mixed-strategy equilibrium.

3. Since the kicker's direction choice will change based on the goalkeeper's dive direction, the kicker's strategy is not a dominant one.

4. (a) 9

		Goalkeeper	
		Right	Left
Shooter	Right	(0,0)	(1,-1)
	Left	(1,-1)	(0,0)

(b) 6

(c) 3

(d) See table below.

		Lee	
		Attack	Wait
Johan	Attack	(3,12)	(9,6)
	Wait	(9,6)	(0,15)

5. (a) Player A will go with Choice 1 $\frac{1}{2}$ of the time and Choice 2 $\frac{1}{2}$ of the time.

   (b) Player B will go with Choice 1 $\frac{2}{5}$ of the time and Choice 2 $\frac{3}{5}$ of the time.

6. (a) Student 1's payoff would be 5, Student 2's payoff would be 4, and Student 3's payoff would be 3.

   (b) Student 1's payoff would be 1, Student 2's payoff would be 0, and Student 3's payoff would be −1.

   (c) Student 1's payoff would be 3, Student 2's payoff would be 3, and Student 3's payoff would be 1.

   (d) Student 1's payoff would be .5, Student 2's payoff would be −.5, Student 3's payoff would be −1.5, and Student 4's payoff would be −2.5.

   (e) Student 1's payoff would be 3.5, Student 2's payoff would be 2.5, Student 3's payoff would be 1.5, and Student 4's payoff would be .5.

7. (a) See table below.

		A	
		H	T
B	H	(0,3)	(2,0)
	T	(2,0)	(0,1)

   (b) The mixed strategy for player B is to play H $\frac{1}{4}$ of the time and T $\frac{3}{4}$ of the time. Player A should play each face half of the time.

(c) The expected winnings for player A for one play is $\frac{6}{8}$ and thus for 16 plays would be \$12. The expected winnings for player B for one play is \$1 and thus for 16 plays would be \$16.

8. See table below.

		Bar 2		
		2	4	5
Bar 1	2	(10000,10000)	(14000,12000)	(14000,15000)
	4	(12000,14000)	(20000,20000)	(28000,15000)
	5	(15000,14000)	(15000,28000)	(25000,25000)

5.4 Solutions

1. If a pilot had to land, it would be helpful to know which airport is closest. A Voronoi diagram could show the pilot the closest airport if the seed points are airports.

2. (a) See figure below.

(b) See figure below.

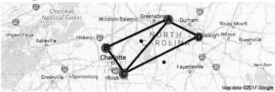

(c) See figure below.

3. (a) See figure below.

(b) If the national park fire fighters start at the lines in the Voronoi diagram separating the cells, they are equidistant from two forest fires and thus can work on either fire from this line.

Index